T0305738

Vibration Engineering

Vibration Engineering presents recent developments in the field of engineering, encompassing industrial norms, applications within the finite element method, infrastructure safety assessment, and active vibration control strategies. It offers a study in seismic vibration control and analysis for building structures and liquid storage tanks.

Spanning across the multiple domains of vibration engineering, this book highlights machinery diagnostics, modal analysis, energy harvesting, balancing, vibration isolation, and human–vibration interaction. It discusses experimental fault identification in journal bearings using vibration-based methods. This book also considers advances in vibration-based structural health monitoring of civil infrastructures.

This book will be a useful reference for industry professionals and engineers facing challenges while dealing with the vibrations in the fields of mechanical, aerospace, structural, and civil engineering.

Vibration Engineering
Modeling, Simulation, Experimentation, and Applications

Edited by
Nitesh P. Yelve and Susmita Naskar

CRC Press
Taylor & Francis Group
Boca Raton London New York

CRC Press is an imprint of the
Taylor & Francis Group, an **informa** business

MATLAB® and Simulink® are trademarks of The MathWorks, Inc. and are used with permission. The MathWorks does not warrant the accuracy of the text or exercises in this book. This book's use or discussion of MATLAB® or Simulink® software or related products does not constitute endorsement or sponsorship by The MathWorks of a particular pedagogical approach or particular use of the MATLAB® and Simulink® software.

Designed cover image: Shutterstock

First edition published 2024
by CRC Press
2385 NW Executive Center Drive, Suite 320, Boca Raton FL 33431

and by CRC Press
4 Park Square, Milton Park, Abingdon, Oxon, OX14 4RN

CRC Press is an imprint of Taylor & Francis Group, LLC

ISBN: 978-1-032-51528-1 (hbk)
ISBN: 978-1-032-51530-4 (pbk)
ISBN: 978-1-003-40269-5 (ebk)

DOI: 10.1201/9781003402695

Typeset in Times
by codeMantra

Contents

Preface...vii

Editors..viii

Contributors ix

Chapter 1   Experimental Fault Identification in Journal Bearings
Using Vibration-Based Condition Monitoring.....................................1

Pallavi Khaire and Vikas Phalle

Chapter 2   Advances in Vibration-Based Structural Health
Monitoring of Civil Infrastructure.....................................12

K. Lakshmi

Chapter 3   3D Vibration Control of Flexible Manipulator Using Inverse
System and Strain Feedback 25

*Minoru Sasaki, Daiki Maeno, Muguro Joseph,
Mizuki Takeda, Waweru Njeri, and Kojiro Matsushita*

Chapter 4   Seismic Vibration Control of a Building Structure
Using a Deep Liquid-Containing Tank with
Submerged Cylindrical Pendulum Appendage 32

Tanmoy Konar and Aparna (Dey) Ghosh

Chapter 5   Seismic Analysis of Base-Isolated Liquid Storage Tanks Using
Supplemental Clutching Inerter 41

*Ketan Narayanrao Bajad, Naqeeb Ul Islam,
and Radhey Shyam Jangid*

Chapter 6   Optimal Piezo Patch Placement Using Genetic Algorithm
and Performance Evaluation of Various Optimal Controllers
for Active Vibration Control of Cantilever Beam 51

Yusuf Khan, S. M. Khot, and Khan Nafees Ahmed

Chapter 7   Design and Fabrication of a Personnel Noise Enclosure for a
Stone-Crushing Unit..62

*Ansaf Mohammed Ashraf, Rohan, P.P., Devsuriya Devan,
Nithin, R., and Sudheesh Kumar, C.P.*

Chapter 8   Investigation of Effect of Porous Material on Performance of
Helmholtz Resonator 71

Nilaj N. Deshmukh, Afzal Ansari, and Axin A. Samuel

Chapter 9   Application of Finite Element Method for Analyzing the
Influence of Geometrical Parameters of Spur Gear Pair
on Dynamic Behavior 84

Achyut. S. Raut, S. M. Khot, and Vishal G. Salunkhe

Index 97

Preface

In recent times, many investigations in the field of vibration have been motivated by its engineering applications. These applications encompass machinery diagnostics; structural health monitoring; automotive noise, vibration, and harshness; human health and biomedical engineering; acoustics; energy harvesting; etc. The authors reviewed many such research works and, after a meticulous review, compiled a collection of novel works in the form of a book. The authors named this book *Vibration Engineering: Modeling, Simulation, Experimentation, and Applications* because they wished to collect and showcase the progress happening in the field of vibration, with respect to its different applications, in one place. This collection of papers in the form of chapters will be useful to nascent as well as expert researchers in academia and industries to understand the level of vibration-related research happening around the world in diverse domains. The authors of this book are thankful to the office bearers at the Council of Vibration Specialists (Dr. Harvindar Singh Gambhir and Dr Tarapada Pyne), faculty members at Fr C R Institute of Technology, Vashi (Dr S M Khot, Dr Nilaj Deshmukh, and Dr Aqleem Siddiqui), and authors of the individual chapters for their support and contribution to this book through INVEST 2022.

MATLAB® is a registered trademark of The MathWorks, Inc. For product information, please contact:
   The MathWorks, Inc.
   3 Apple Hill Drive
   Natick, MA 01760-2098 USA
   Tel: 508-647-7000
   Fax: 508-647-7001
   E-mail: info@mathworks.com
   Web: www.mathworks.com

Prof Nitesh P. Yelve

Prof Susmita Naskar

Editors

Prof. Nitesh P. Yelve is presently working as an Assistant Professor in the Department of Mechanical Engineering of the Indian Institute of Technology (IIT) Bombay, India. He received his PhD from the Department of Aerospace Engineering of IIT Bombay. He also pursued the Postdoctoral Fellowship at the City University of Hong Kong. The areas of his research interest are structural health monitoring using ultrasonic guided waves and vibration methods, active vibration control, structural dynamics, and composite materials. He has presented 51 papers at national and international conferences in India and abroad, and published 32 papers in international journals. He is a Chartered Engineer and Certified Level II Vibration Analyst. He is a Fellow of The Institution of Engineers (India) and the Council of Vibration Specialists. He is also a Member of ASME and a Senior Member of IEEE and IIAV.

Prof. Susmita Naskar is an Assistant Professor in the Faculty of Engineering & Physical Sciences at the University of Southampton. She worked as a postdoctoral research fellow at the Whiting School of Engineering of Johns Hopkins University in the collaboration of Army Research Lab, USA. She moved to the USA after completing her doctoral degree from the University of Aberdeen.

Susmita's research interests and expertise broadly lie in the field of multi-scale structural mechanics and multi-physics analysis focusing on engineered materials and advanced composites involving the intersection of additive manufacturing, material characterization through computational design, and experiments in engineering. In addition, she is also working in advanced manufacturing techniques that are relevant to the fabrication of engineered materials like composites and metamaterial.

Contributors

Khan Nafees Ahmed
Department of Automobile Engineering,
  Anjuman-I-Islam's M. H. Saboo
  Siddik College of Engineering
  Byculla, Mumbai

Afzal Ansari
Department of Mechanical Engineering,
  Agnel Charities' Fr. C. Rodrigues
  Institute of Technology, Vashi, Navi
  Mumbai, India

Ansaf Mohammed Ashraf
Department of Mechanical Engineering,
  Government College of Engineering,
  Kannur, Kerala, India

Ketan Narayanrao Bajad
Department of Civil Engineering,
  Indian Institute of Technology (IIT)
  Bombay, Powai, Mumbai, India

Nilaj N. Deshmukh
Department of Mechanical Engineering,
  Agnel Charities' Fr. C. Rodrigues
  Institute of Technology, Vashi, Navi
  Mumbai, India

Devsuriya Devan
Department of Mechanical Engineering,
  Government College of Engineering,
  Kannur, Kerala, India

Aparna (Dey) Ghosh
Department of Civil Engineering,
  IIEST, Shibpur, Howrah, India

Naqeeb Ul Islam
Department of Civil Engineering,
  Indian Institute of Technology (IIT)
  Bombay, Powai, Mumbai, India

Radhey Shyam Jangid
Department of Civil Engineering,
  Indian Institute of Technology (IIT)
  Bombay, Powai, Mumbai, India

Muguro Joseph
Intelligent Production Technology
  Research & Development Center
  for Aerospace, Gifu University,
  Yanagido, Gifu, India
School of Engineering, Dedan Kimathi
  University of Technology, Nyeri,
  Kenya

Pallavi Khaire
Department of Mechanical Engineering,
  Veermata Jijabai Technological
  Institute, Mumbai, India
Department of Mechanical Engineering,
  Fr. C. Rodrigues Institute of
  Technology, Vashi, India

Yusuf Khan
Department of Mechanical Engineering,
  Fr. Conceicao Rodrigues Institute of
  Technology, Vashi, India
Department of Mechanical Engineering,
  Anjuman-I-Islam's Kalsekar
  Technical Campus, New Panvel,
  Mumbai

S. M. Khot
Department of Mechanical Engineering,
  Fr. Conceicao Rodrigues Institute of
  Technology, Vashi, Navi Mumbai,
  India

Tanmoy Konar
Department of Civil Engineering,
  IIEST, Shibpur, Howrah, India

Sudheesh Kumar, C.P.
Department of Mechanical Engineering,
Government College of Engineering,
Kannur, Kerala, India

K. Lakshmi
Structural Health Monitoring
Laboratory, CSIR-Structural
Engineering Research Centre,
CSIR Campus, Taramani, Chennai,
Tamilnadu, India

Daiki Maeno
Department of Mechanical Engineering,
Faculty of Engineering, Gifu
University, Yanagido, Gifu, Japan

Kojiro Matsushita
Department of Mechanical Engineering,
Gifu University, Yanagido, Gifu,
Japan

Nithin, R.
Department of Mechanical Engineering,
Government College of Engineering,
Kannur, Kerala, India

Waweru Njeri
Department of Electrical and Electronic
Engineering, School of Engineering,
Dedan Kimathi University of
Technology, Nyeri, Kenya

Vikas Phalle
Department of Mechanical Engineering,
Veermata Jijabai Technological
Institute, Mumbai, India

Achyut S. Raut
Department of Mechanical Engineering,
Rajendra Mane College of
Engineering and Technology,
Ambav, India
Department of Mechanical Engineering,
Fr. C. Rodrigues Institute of
Technology, Vashi, Navi Mumbai,
India

Rohan, P.P.
Department of Mechanical Engineering,
Government College of Engineering,
Kannur, Kerala, India

Vishal G. Salunkhe
Department of Mechanical Engineering,
Fr. C. Rodrigues Institute of
Technology, Vashi, Navi Mumbai,
India

Axin A. Samuel
Department of Mechanical Engineering,
Agnel Charities' Fr. C. Rodrigues
Institute of Technology, Vashi, Navi
Mumbai, India

Minoru Sasaki
Intelligent Production Technology
Research & Development Center for
Aerospace, Gifu University, Tokai
National Higher Education and
Research System, Yanagido, Gifu,
Japan

Mizuki Takeda
Department of Mechanical Engineering,
Faculty of Engineering, Gifu
University, Yanagido, Gifu, Japan

1 Experimental Fault Identification in Journal Bearings Using Vibration-Based Condition Monitoring

Pallavi Khaire
Veermata Jijabai Technological Institute
Fr. C. Rodrigues Institute of Technology

Vikas Phalle
Veermata Jijabai Technological Institute

1.1 INTRODUCTION

Hydrodynamically lubricated journal bearings are commonly used in machinery like turbines because of their excellent durability and good load-bearing capacity. That's why they are one of the essential machine parts for supporting the rotating machinery shaft. As the performance of journal bearing is characterized by various parameters, the design engineers generally choose these parameters from an analysis of bearing characteristics. Sometimes use of these bearings is critical, and their failure may cause the function of the entire plant to come to a halt, which may lead to a large production loss and even a catastrophe. Therefore, the maintenance of journal bearings becomes very important. For very critical applications, an on-time maintenance process is used called Condition Monitoring. Condition monitoring helps predict the lead time to failure and avoids a sudden failure of the machine [1]. Journal bearings are commonly used for supporting the shaft in a rotating machinery which bears heavy loads. The most common issues that can arise are fatigue failure and failure due to corrosion, which increase safety risks. Thus, the foremost requirement is to choose an efficient condition-monitoring technique for identifying the faults at the earliest to avoid sudden breakdowns.

Researchers have carried out extensive research to identify faults in journal bearings using various condition-monitoring techniques. Mathew et al. [2] presented an exhaustive review of different methods of condition-based monitoring including the procedure and testing summary of vibration signatures for detecting upcoming failure. Shi et al. [3] proposed a different approach that involves a

DOI: 10.1201/9781003402695-1

combination of wavelet transform and envelop spectrum for calculating and select-ing statistical features that are influenced by defects. Earlier, many time-frequency analysis techniques used to choose statistical features for faults in bearing based on vibration signals had no means to obtain high time and frequency resolutions. Lu et al. [4] introduced a new method called Sound-Aided Vibration Signal Adaptive Stochastic Resonance (SAVASR) to effectively improve weaker signals which are repetitive with respect to time, fully immersed in heavy background noise ratio (SNR) which cannot be enhanced by traditional ASR method on a bearing setup where only vibrational signal was processed. The success rate of SAVASR is higher in detecting correct frequency components under low SNR conditions. Therefore, this process has an efficient application for detecting bearing defects automatically. Surojit Poddar et al. [5] carried out a research to monitor the effect of externally injected contaminants in journal-bearing oil using acoustic emission (AE) and vibration monitoring techniques. Antoni et al. [6] conducted an experi-ment on fault detection of roller element bearings, in which strong interfering gear signals were present, which is mostly observed in gearboxes of helicopters. It was observed that gear signals were completely periodic, but bearing signals were seen randomly.

Narendiranath et al. [7] proposed a study on fault detection in journal bearing using Debauchies Wavelet (db02) with various defect conditions such as half loose-ness, groove, hole, indentations, and full looseness. A test setup was prepared to capture vibration data in the form of time domain and frequency domain using FFT analyzer. The signal decomposition using wavelet transform was carried out in MATLAB. The fast Fourier transform was selected to get the frequency spectrum where the highest peak of amplitude was observed at 6× of the operating frequency for defective journal and at 3× for looseness. To study the anomalies, various operat-ing conditions are used in the journal bearing such as full quantity of lubricant, half looseness, half quantity of lubricant, groove, hole, indentations, and full loose. It was concluded that Artificial Neural Networks (ANNs) are very effective in classifying faults. Parno et al. [8] concluded that fault diagnosis plays a prime role in prevent-ing major losses to industries through condition-monitoring technique. Premature failure often occurs in journal bearings due to the presence of fluid and solid con-taminants which can cause a mechanical component to generate surface vibration. Excessive vibrations cause the journal bearings to fail, which can lead to monetary losses and crucial safety issues; hence, monitoring them and diagnosing any prob-lems as early as possible are required. Jung et al. [9] found that failures of journal bearings are mainly linked to mechanical instabilities produced by lubricant-related system problems and worn-out journal bearing, which produces high amplitude of shaft vibration. Saridakis et al. [10] introduced a new fault diagnosis model that uses ANNs for identifying the increase in wear severity and/or the increment of the angle of misaligned shaft.

A lot of research has been carried out for the diagnosis of rolling element bearing defects using vibration analysis, but the work related to vibration-based condition monitoring of journal bearings is attempted by very few researchers. Therefore, in this study, an attempt is made to identify the unique vibration characteristics in the vibration spectrum that occurred due to different journal-bearing defects.

1.2   METHODOLOGY

All the test rig shown in Figure 1.1 has a 0.25 HP DC motor with driver and driven shaft, a flexible flange coupling, C channels, and a pulley and rope system to lift the main shaft and journal bearings. Journal bearings provide support for the shaft, and one of them will be used as a test bearing whose vibration signatures will be captured with a tri-axial accelerometer. The length of the shaft is 600 mm, and the bearings are placed 300 mm apart. The MS shaft has 25.2 mm diameter. All the elements of the setup, i.e., journal bearings, motor, and pulley, are fastened on C channels of appropriate sizes. The C channels are welded on a metal base of size 700 mm × 300 mm × 6 mm. The specifications of the journal bearing are mentioned in Table 1.1. The metal base is fastened to a wooden base with counter-sunk bolts. A rubber sheet is placed between the metal and wood base. This is done for vibration damping of the setup as the shaft rotates at a high speed and also to maintain

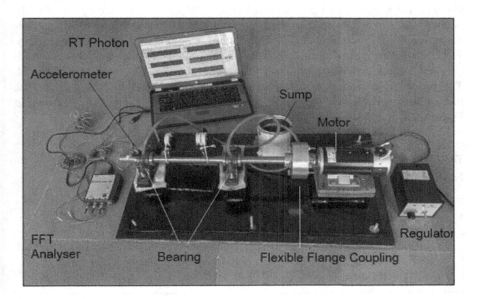

FIGURE 1.1   Experimental test rig.

TABLE 1.1
Journal-Bearing Specifications

Sr. No.	Parameter
1	Length of bearing, $L = 50$ mm
2	Diameter of bearing, $D = 25.4$ mm
3	Diametrical clearance, $C = 0.2$ mm
4	Shaft diameter $= 25.2$ mm
5	$L/D = 1.96$

the vertical alignment of the setup. Before starting the motor, the alignment of the shaft is inspected using dial gauges. Also, the surface level is checked using the spirit level. Initially, a DC pump was placed in the oil container for pumping and supplying the oil to two journal bearings with a 0.25-inch diameter pipe. Then, the motor is started to drive the main shaft. Initially, the test rig is run for 1–2 minutes to settle all the minor vibrations. The accelerometer is mounted on top of the housing and connected with the FFT analyzer. After checking the alignments, the speed of the motor is increased slowly with the help of a speed regulator. When the journal (shaft) just starts to rotate, load is given to the pulley placed on a C channel within the bearing span in order to lift the shaft. This is done to create hydrodynamic action in the bearings so that an oil film responsible for bearing action is generated. As the oil film is generated, the vibration signatures are captured on the bearing on which the accelerometer is placed.

A tri-axial, piezoelectric accelerometer is used along with the RT Photon+ (4326 A, B&K make) FFT analyzer. It has four channels and a sensitivity of 0.316 pC/ms². The time domain data are processed and displayed at a computer terminal through RT Photon+ interface. Signals are acquired in the frequency domain. Then, the healthy bearing on which the accelerometer is placed is changed with bearings having different faults, and their vibration data are stored and observed. This procedure is carried out for all the given faults (mentioned in Table 1.2), and the plots of all the respective faults are recorded. The defects considered are shown in Figures 1.2a,b and 1.3.

TABLE 1.2
Faults Created on the Bearing

Sr. No.	System	Description
1	Healthy Bearing System	The system is without any defects.
2	Half Looseness Defect	The housing bolts of the bearing are slightly loosened.
3	Full Looseness Defect	The housing bolts of the bearing are completely loosened.
4	Groove Defect	A groove is made at the inner contacting surface.
5	Hole Defect	A hole is drilled at the inner contacting surface.

FIGURE 1.2   (a) Groove defect. (b) Hole defect.

FIGURE 1.3 Looseness defect.

A groove on the bearing was created using welding, and a hole defect was created by the chemical etching process. The half looseness was achieved by loosening the left nut of the bearing housing, and for full looseness, both left and right nuts were loosened. The setup was tested with the accelerometer placed above the bearing head.

1.3   RESULTS

In the experiment conducted, a series of readings in the frequency domain were taken for a healthy system at speeds 900, 1200, and 1500 rpm. The vibration spectrum and the nature of each spectrum can be analyzed from Figures 1.4, 1.5, and 1.6 for 900, 1200, and 1500 rpm, respectively.

In Figure 1.4, as the shaft is rotating at 900 rpm, the fundamental frequency is 900/60 = 15 Hz. The significant peak can be seen at 1× of the fundamental frequency, i.e., at 15 Hz. In Figure 1.5, as the shaft is rotating at 1200 rpm, the fundamental frequency is 1200/60 = 20 Hz. Thus, the significant peak can be seen at 1× of the fundamental frequency, i.e., at 20 Hz.

In Figure 1.6, as the shaft is rotating at 1500 rpm, the fundamental frequency is 1500/60 = 25 Hz. Thus, the significant peak can be seen at 1× of the fundamental frequency, i.e., at 25 Hz. From the frequency domain plots of healthy bearing, it can be observed that the significant peak can be seen at 1× frequency for all the speeds. Thus, it can be inferred that peak is obtained at 1× of the fundamental frequency for a healthy system irrespective of its speed. Furthermore, the system with different faults was tested at 900, 1200, and 1500 rpm. The vibration signature plots at 1200 rpm for different fault conditions such as half looseness, full looseness, groove, and hole can be observed from Figures 1.7, 1.8, 1.9, and Figure 1.10, respectively. For a system with half looseness, the significant peak is obtained at 3× of the fundamental frequency as shown in Figure 1.7, and a system with full looseness shows the

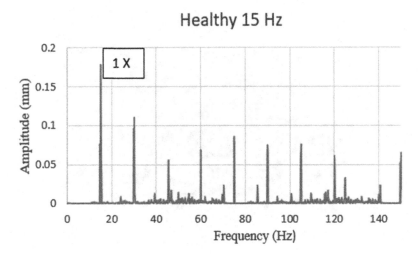

FIGURE 1.4   Vibration spectrum of healthy system at 900 rpm.

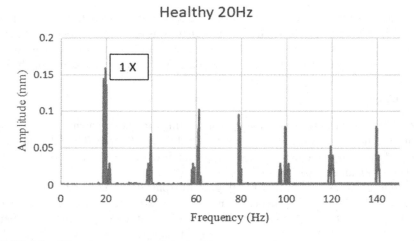

FIGURE 1.5   Vibration spectrum of the health system at 1200 rpm.

significant peak at 3× of the fundamental frequency as observed in Figure 1.8. The vibration amplitude for full looseness is almost twice as compared to the amplitude observed for half looseness.

For unhealthy system with bearing having groove on the journal surface, the significant peak is obtained at 6× of the fundamental frequency as depicted in Figure 1.9. For unhealthy system with bearing having a hole on the inner surface, the significant peak is obtained at 6× of the fundamental frequency as shown in Figure 1.10.

It can be observed from Figure 1.11 that the maximum amplitude for a healthy system is seen at 1× of the shaft speed. For looseness, it shifted to 3×, and for groove and hole, it is at 6× of the operating speed. These unique vibration characteristics

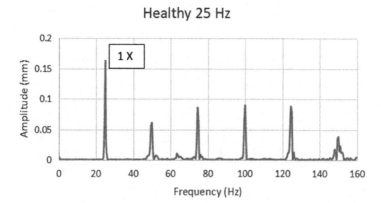

FIGURE 1.6   Vibration spectrum of the healthy system at 1500 rpm.

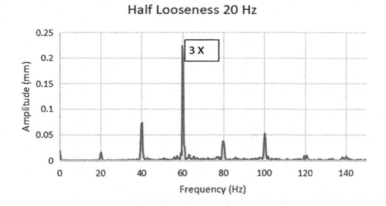

FIGURE 1.7   Vibration spectrum of half looseness defect at 1200 rpm.

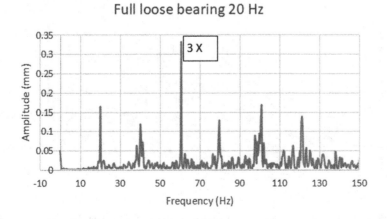

FIGURE 1.8   Vibration spectrum of full looseness defect at 1200 rpm.

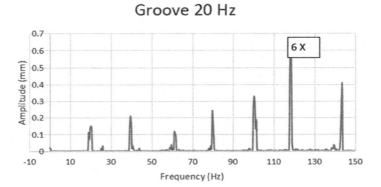

FIGURE 1.9   Vibration spectrum of groove defect at 1200 rpm.

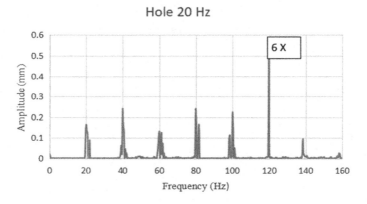

FIGURE 1.10   Vibration spectrum of hole defect at 1200 rpm.

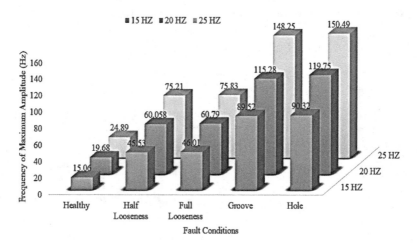

FIGURE 1.11   Comparative fault frequencies of healthy and unhealthy systems at different operating speeds.

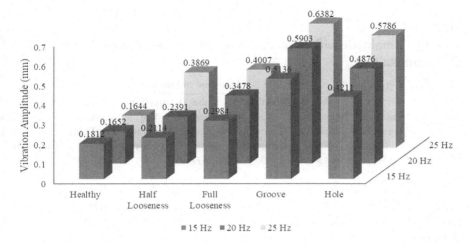

FIGURE 1.12 Comparative amplitude of vibration for healthy and unhealthy systems at different operating speeds.

are independent of shaft speed and severity. There is a slight frequency fluctuation noticed because of the fluctuation in shaft speed. The amplitude of vibration increases as the operating speed increases, which can be clearly observed in Figure 1.12. The level of vibration is more in the case of groove defect. Table 1.3 shows vibration amplitudes with respect to operating speeds in all three directions. Along the axial direction, the distinctive vibration characteristics of the defect were unseen. In the radial direction, the maximum amplitude of vibration was observed at the fault frequencies as reported in the literature.

1.4 CONCLUSION

In fault detection by condition monitoring using the vibration analysis approach, parameters in the frequency spectrum are more consistent than time domain parameters in the detection of a fault. The frequency of significant peaks does not change with the speed as the peaks are obtained at $1\times$ frequency for all three speeds of 900, 1200, and 1500 rpm for healthy bearing. The phase angle associated with a healthy system was observed as 0°. The maximum amplitude of vibration for half looseness and full looseness is seen at $3\times$ frequency. But the amplitude value for full looseness is almost two times more than the maximum amplitude for half looseness. Along with the maximum amplitude, side bands were also noticed. The spacing between sidebands was found to be $2\times$. The phase angle obtained is 120° for half looseness, and for full looseness, it is 150°. The peaks obtained for hole defect have considerably high amplitudes than those of healthy bearing, and a significant peak is obtained at $6\times$ of the operating frequency. The phase plot obtained is 90° for the hole and 180° for the groove. The unique vibration characteristics observed for each fault remain unchanged irrespective of operating speed. The value of peak amplitude in a system goes on decreasing in the following order: groove, hole, full looseness,

TABLE 1.3

Vibration Amplitude with Respect to Speed in Three Different Directions (Axial and Radial)

		Direction of Measurement					
		Axial -X Direction		Radial -Y Direction		Radial -Z Direction	
Defect	Operating Frequency (Hz)	Maximum Amplitude (mm)	X of Operating Frequency	Maximum Amplitude (mm)	X of Operating Frequency	Maximum Amplitude (mm)	X of Operating Frequency
No Defect (Healthy)	15 Hz	0.1543	1X	0.1577	1X	0.1812	1X
	20 Hz	0.1612	1X	0.1532	1X	0.1652	1X
	25 Hz	0.1827	1X	0.1735	1X	0.1644	1X
Half Looseness	15 Hz	0.2031	1X	0.2172	3X	0.2114	3X
	20 Hz	0.2109	1X	0.2651	3X	0.2391	3X
	25 Hz	0.1969	1X	0.3319	3X	0.3869	3X
Full Looseness	15 Hz	0.3167	1X	0.3176	3X	0.2984	3X
	20 Hz	0.3412	1X	0.3645	3X	0.3478	3X
	25 Hz	0.3689	1X	0.3788	3X	0.4007	3X
Groove	15 Hz	0.3554	1X	0.4445	6X	0.5136	6X
	20 Hz	0.4101	1X	0.5302	6X	0.5903	6X
	25 Hz	0.4302	1X	0.5542	6X	0.6382	6X
Hole	15 Hz	0.3116	1X	0.3835	6X	0.4211	6X
	20 Hz	0.3418	1X	0.4347	6X	0.4876	6X
	25 Hz	0.3478	1X	0.5132	6X	0.5786	6X

half looseness, and healthy bearing. The phase plot allows distinguishing among these faults when the significant peaks are obtained at the same multiples of frequency. The vibration characteristics for the considered defects are matching with the vibration characteristics mentioned by researchers in the literature.

1.5 FUTURE SCOPE

As far as the future scope of this project is concerned, the vibration data can be fed to the machine learning algorithm which will identify the fault in the system automatically. More defects can also be created like oil whirl, and indentations on the bearing surface and their vibration spectra can be studied and identified on the basis of a literature review. These spectra can thus be used to train the algorithm. An attempt to integrate the real-time machine learning system with on-time condition monitoring can also be done. Moreover, the machine learning program can be interfaced with the mobile app using an Internet of Things (IOT)–based system.

ACKNOWLEDGMENTS

We want to express our deepest gratitude to all those who have provided us with the opportunity and possibility to complete this research. A special thanks go to our institutes, VJTI, FCRIT Vashi, and CoE for providing us with the research facility and motivating us to pursue research in the machine learning domain.

REFERENCES

[1] Sheng, C., Li, Z., Qin, L., Guo, Z., and Zhang, Y. (2011). Recent progress on mechanical condition monitoring and fault diagnosis. *Proced. Eng.* 5:142–146.

[2] Mathew, J., and Alfredson, R. J. (1984). The condition monitoring of rolling element bearings using vibration analysis. *J. Vib. Acoust. Stress. Reliab.* 106(3):447–453.

[3] Shi, D. F., Wang, W. J., and Qu, L. S. (2004). Defect detection for bearings using envelope spectra of wavelet transform. *ASME. J. Vib. Acoust.* 126(4):567–573.

[4] Siliang, L., Ping, Z., Yongbin, L., Zheng, C., Hui, Y., and Qunjing, W. (2019). Sound-aided vibration weak signal enhancement for bearing fault detection by using adaptive stochastic resonance. *J. Sound. Vib.* 449:18–29. https://doi.org/10.1016/j.jsv.2019.02.028

[5] Poddar, S., and Tandon, N. (2019). Detection of particle contamination in journal bearing using acoustic emission and vibration monitoring techniques. *Tribol. Int.* 134:154–164.

[6] Antoni, J., and Randall, R. B. (2002). Differential diagnosis of gear and bearing faults. *ASME. J. Vib. Acoust.* 124(2):165–171.

[7] Narendiranath, B. T. (2017). Journal bearing fault detection based on daubechies wavelet. *Archiv. Acoustics.* 42(3):401–414.

[8] https://shodhganga.inflibnet.ac.in/bitstream/10603/142234/13/13_chapter%205.pdf.

[9] Jung, J., Park, Y., Lee, S., Cho, C., Kim, K., Wiedenbrug, E., and Teska, M. (2015). Monitoring of journal bearing faults based on motor current signature analysis for induction motors. In *Proceedings of IEEE Energy Congress and Exposition*, Montreal, QC, 300–307. DOI:10.1109/ECCE.2015.7309702

[10] Saridakis, K., Nikolakopoulos, P., Papadopoulos, C., and Dentsoras, A. (2008). Fault diagnosis of journal bearings based on artificial neural networks and measurements of bearing performance characteristics. In B.H.V. Topping and M. Papaddrakakis (Eds), *Proceedings of the Ninth International Conference on Computational Structures Technology*, Civil-Comp Press, Stirlingshire, Paper 118. https://doi.org/10.4203/ccp.88.118

2 Advances in Vibration-Based Structural Health Monitoring of Civil Infrastructure

K. Lakshmi
CSIR-Structural Engineering Research Centre

2.1 INTRODUCTION

During the past two decades, structural health monitoring (SHM) has gained importance as a potential research domain for civil engineering. In particular, the focus of the researchers has increased in developing the vibration-based damage detection techniques, exploiting the output-only responses with the help of efficient algorithms. An ample literature is available on SHM systems dealing with the damage diagnostics of structures [1]. Most of the damage diagnostic techniques are developed by tracking the changes in dynamic characteristics such as natural frequency, mode shape curvature, modal flexibility, and modal strain energy. Even though these modal-based approaches are very appealing on paper, several practical difficulties exist in moving them for practical structures. Some of the major issues with them are as follows: (i) extracting higher modes where the damage features are present is a difficult proposition especially with ambient loads, (ii) modes are insensitive to subtle cracks, and (iii) dealing with environmental variability is extremely challenging with modal-based approaches. So, several other vibration-based techniques like multivariate analysis, time-frequency analysis, and time series analysis were gaining popularity in the past decade. Among them, time series models are being popularly used in the recent past [2,3]. Time series models are preferred as they are model-free and can be used in wireless sensor networks for real-time monitoring.

Motivated by the advantages of time series models and advanced signal processing techniques, it is proposed to seek solutions to the issues on vibration-based damage diagnosis by employing multi-model strategies. The recent research advances in the field of vibration-based techniques for SHM using multi-model strategies for (i) detection of minor damages and (ii) handling EOV are presented in this chapter. Innovative solutions to isolate the interesting components of the signal by decomposition techniques for enhancing the efficiency of time series methods are presented.

12

DOI: 10.1201/9781003402695-2

Also, as a cost-effective alternative to the traditional SHM, indirect SHM, using indirect measurements from an instrumented vehicle moving over the bridge, is gaining popularity in the recent past for damage diagnosis. SHM using these indirect measurements has several advantages over the more conventional direct approach with a fully instrumented bridge. Even though only a single sensor is used to collect the time-history data in the indirect approach, it offers much higher spatial resolution than the traditional direct approach. This is because the instrumented vehicle in the indirect approach can receive the vibration characteristics of each point as it passes over the bridge. This chapter also presents a signal processing–based modal identification technique using indirect SHM by vehicle scanning.

2.2   DETECTION OF SUBTLE OPEN CRACKS IN THE STRUCTURE

The prime objective here is to detect the cracks even at their incipient condition so that timely repair at the damage initiation stage avoids major costs incurred while attending the damage to the structure that is identified at the much more advanced stage. The responses of the structure with subtle open cracks are basically linear. However, the alterations in the dynamic traits due to subtle cracks are minor and are often masked in the measurement noise. It is extremely difficult to diagnose these subtle cracks using raw dynamic signatures and any of the available current damage diagnostic techniques. Due to this, it is proposed to decompose the raw time-history signals measured using the sensors spatially located on the structure. The decomposed signals which have rich damage features are isolated and reconstructed to perform damage diagnostics.

In this chapter, the performance of the multi-model damage diagnostic techniques is demonstrated by considering subtle open cracks. Signal decomposition is performed using the following techniques: (i) Singular Spectrum Analysis (SSA), (ii) Blind Source Separation (BSS), and (iii) Empirical Mode Decomposition (EMD), which are popularly being used in several other fields. Numerical simulation studies and experimental verification on a simply supported beam simulating open cracks are performed. In this chapter, experimental studies are presented to demonstrate the practical application of the proposed techniques.

2.2.1   Singular Spectrum Analysis–Based Multi-model Technique

The fundamental goal of SSA is to break up the realized time series into constituent components such as trend, periodic, seasonal, and noise [4]. This is achieved in SSA through an Eigen decomposition of trajectory matrix. The first step is "Embedding", which is to form a Hankel matrix by sliding a window on a realized time series. The length of the window is required to be shorter than the original series. Every position of the sliding window gives rise to a column in the Hankel matrix. The Hankel matrix, in the next step of "Singular Value Decomposition (SVD)", is split into a group of elementary matrices with descending norm. In the next step, "Grouping", an approximation of the Hankel matrix is obtained by curtailing the addition of elementary matrices. In the process of grouping, the elementary matrices that do not signify to the original matrix norm are ignored. In this way, the Hankel matrix is converted into an approximated time series, averaging the diagonals and smoothed. This process is termed as "Diagonal Averaging" or "Reconstruction". The mathematical formulations related to SSA can be found elsewhere [4,5]. Further, SSA is sensitive to window length. The automated procedures for selection of window length and selection of signals for reconstruction of the time series with damage-rich features are discussed in Lakshmi et al. [5]. The details of the damage diagnosis process using SSA and AutoRegressive Moving Average with eXogenous input (ARMAX) time series model are shown schematically in Figure 2.1.

2.2.2 BLIND SOURCE SEPARATION (BSS)–BASED MULTI-MODEL TECHNIQUE

A multi-model damage detection technique combining improved Second-Order Blind Identification (SOBI) [6] and time series analysis to detect the minor incipient damage in the structural system is proposed. The technique is a baseline-free, output-only diagnostic method that analyses the acceleration time-history data from the interested structure. The acceleration data subsets are preprocessed by Discrete Wavelet Transform (DWT), using Daubechies wavelet db4 to visualize the discontinuities in the signal more clearly. The wavelet domain signals obtained on a selected scale, for the baseline as well as the current data subsets, are considered as the mixtures and are fed into the BSS model. Improved SOBI algorithm with selection of time lags randomly, used for Joint Approximate Diagonalization (JAD), is used to solve the BSS problem and thereby the mixing matrix is obtained. The sources are then estimated based on the input measurements for all the subsets of the baseline data. Once the sources and mixing matrix are obtained for both the subsets, they are used for reconstructing the new baseline and current subset. After reconstruction, the cepstral

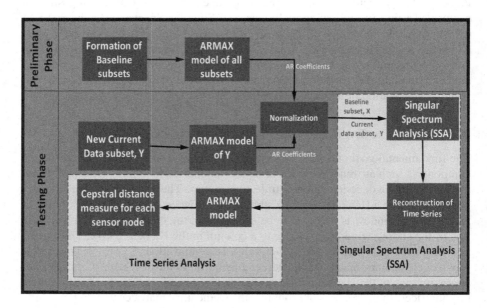

FIGURE 2.1 Damage diagnostic technique combining SSA with the ARMAX model.

distances between the two ARMAX time series models are calculated. The cepstral distance is evaluated as the weighted Euclidean distance of the Cepstrums of any two ARMAX models. Cepstral distances thus obtained for the sensor nodes in a subset are explored for locating the damage. Based on the hypothesis that the cepstral distance peaks up proportional to the difference between any two signals, their magnitudes are verified. The higher magnitude of the cepstral distance measure at a sensor node indicates the existence of damage in the vicinity of that node. The corresponding index number of the subset can be used to evaluate the time instant of damage. Thus, the proposed damage detection methodology using BSS and ARMAX models, with the cepstral damage index [7], can act as a promising tool.

2.2.3 EMPIRICAL MODE DECOMPOSITION (EMD)– BASED MULTI-MODEL TECHNIQUE

EMD [8] is being popularly used for decomposition of time series signals into unimodal signals for performing Hilbert transform. EMD breaks up the dynamic signals from the structure in ambient conditions into Intrinsic Mode Functions (IMFs) using the sifting process, and is used to reconstruct a new time series to construct the ARMAX model for damage diagnosis. A process known as shifting is adopted in EMD. If $x(t)$ is the measured time history, the upper and lower envelopes of $x(t)$ are created by joining the local maxima and minima of $x(t)$, using a cubic spline function. Next, the original signal is subtracted by the mean of the two envelopes. The shifting process is repeated till all the IMFs are extracted from the signal. In this study, a combination of HHT and time series models is used for damage diagnosis to identify the exact time instant of damage occurrence and also spatial location of damage precisely. The vibration response due to ambient loads is obtained as a continuous time-history data. For this purpose, first, EMD is performed using the proposed empirical mode decomposition algorithm with intermittency and an appropriate number of IMFs are chosen, based on the criteria discussed below for reconstruction of the time series.

Instantaneous envelopment ($EV(t)$) is computed after subjecting the IMF, $C(t)$, to Hilbert transform. The strength of the signal of every IMF is evaluated by the following Signal Relative Strength Ratio (SRSR) [9]:

$$SRSR(t) = \frac{\max\{EV(t)\} - \min\{EV(t)\}}{f(t)\left(\sqrt{\sum_{t_i=1}^{N}\{EV(t_i)\}^2}\right)} \qquad (2.1)$$

where the maximum value of $EV(t)$ is denoted as *max* and the minimum value of $EV(t)$ is denoted as *min*. A judiciously chosen tolerance level can be specified, above which if SRSR value of an IMF lies, can be considered as an interesting IMF. Using such chosen IMFs, the time series is reconstructed and time series analysis is performed to identify the spatial location of damage using a predefined damage index evaluation.

2.2.4 EXPERIMENTAL INVESTIGATIONS

Extensive numerical and experimental studies are performed to evaluate the proposed damage diagnostic algorithms for subtle damage detection. However, for the paucity of space, only experimental investigation to prove the performance of the multi-model-based damage detection algorithms to locate the incipient damage such as open cracks is presented in this chapter. An RCC, simply supported beam of length 3 m, width 165 mm, and depth 200 mm, with multiple cracks, is considered for the experimental studies as shown in Figure 2.2. The cracks are induced at 1/3 and 2/3 lengths, through static loading by a hydraulic jack.

Sixteen micro electro mechanical systems accelerometers having a measurement range of ±2 g and a frequency range of 0–350 Hz are mounted at equally spaced locations to record the acceleration time-history data. A modal shaker (with sine peak force capacity 200 N) is used at the center of the beam to excite the random loads, similar to the ambient vibration. Small masses of varying sizes are placed arbitrarily on the specimen to simulate the effect of environmental variability. The position and the size of the masses are changed for each trial of experiment. Plot showing the singular values of SSA and also the location of damage using the cepstral distance measure using the ARMAX model is shown in Figure 2.3a and b, respectively.

FIGURE 2.2 Experimental investigation on simply supported RCC beam with multiple cracks.

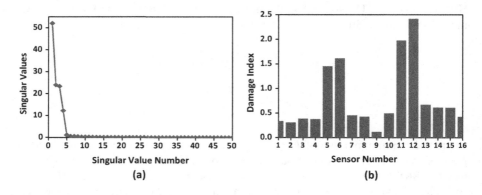

FIGURE 2.3 Multiple damage scenario: (a) components of SSA and (b) damage indices using the SSA-ARMAX model.

It can be seen from Figure 2.3a that the first four SSA components can be grouped to isolate the useful components of trend and cyclic information. The other components are noise related and so are neglected. The grouped components are used for reconstructing the signal, which is then subjected to time series analysis using the ARMAX model. The damage index evaluated using cepstral distance metric is shown in Figure 2.3b. From the figure, it can be observed that the damage indices are high at 6th and 12th elements, which are at actual 1/3 and 2/3 lengths of the beam.

In SOBI-based technique, the acceleration responses related to healthy and damage datasets are processed through wavelet transform with a suitable scale, where the spike is noted in the wavelet coefficients, which is considered as an indication of the presence. The corresponding time instant of damage is evaluated. These wavelet coefficients are subjected to SOBI, and the sources with the sharp spikes are isolated. The isolated sources of the scenarios can be found in Figure 2.4a. The extracted sources are used for reconstruction and for further analysis using time series model. The damage indices using cepstral distances are shown for each sensor in Figure 2.4b, indicating the location of damage precisely.

For EMD-based technique, the cross-correlated signals of the measured acceleration time history are provided as inputs to the EMD handling intermittency algorithm to evaluate the IMFs. The extracted IMFs are shown in Figure 2.5a. The reconstructed signals using the selected IMFs of the signals from current and healthy states are further used in ARMAX time series model. The ARMAX models constructed further are evaluated for the cepstral distances, to locate the damage as shown in Figure 2.5b. The cepstral distance damage indices of ARMAX, for the multiple damage scenario, clearly affirm the damage locations (at 1/3 and 2/3 lengths of the beam).

From Figures 2.3–2.5, it can be seen that the multi-model techniques are efficient in identifying the incipient damage such as subtle cracks even in the presence of environmental variability and measurement noises.

2.3 HANDLING EOV THROUGH A ROBUST TECHNIQUE FOR ONLINE SHM

Early damage detection is the paramount need of an SHM system and its practical field implementation. In particular, a minor damage in its initiation stage brings out only slight changes in the dynamic properties of the structure, thereby altering only few modal responses. Therefore, the damage features of those few modal responses will not be exhibited much in the measured vibration response. Additionally, the existence of environmental variability (EOV), which generally changes the dynamic characteristics of the signal, masks the effect of the minor incipient damage, under conventional damage diagnostic algorithms. To circumvent this, a novel online health monitoring technique to handle the EOV and simultaneously locate the minor damage is proposed. Cointegration technique is employed to handle the EOV in this proposed method. The cointegrating vectors of healthy data, named as 'baseline data', collected from the structure are formed initially. Later, these vectors are used as filters on the hidden influence of EOV of the current data. BSS technique is then employed on the cointegrated time series, cleared from EOV, to obtain the modal responses.

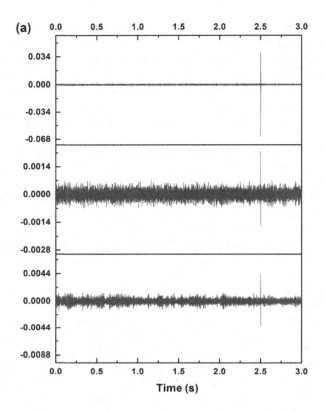

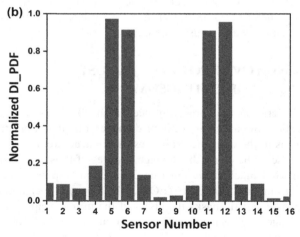

FIGURE 2.4 Multiple damage scenario of RCC beam with damage at 2.5 s: (a) wavelet processed sources showing spikes and (b) normalized cepstral distance measure of the reconstructed current data.

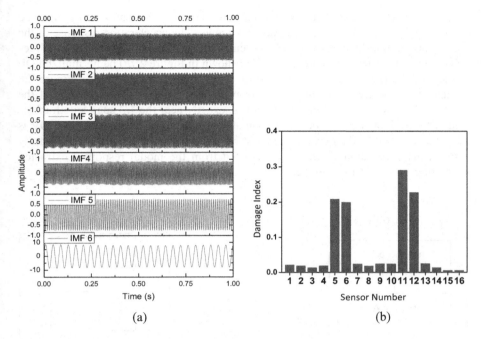

Time (s)   Sensor Number

(a)   (b)

FIGURE 2.5 Results obtained from the EMD-ARMAX model for experimental beam: (a) extracted IMF (b) multiple damage locations.

For M modes from N sensors, the correlation coefficients $\gamma_j[i]$ ($i=1, 2, 3, \ldots$ N; $j=1, 2, \ldots$ M) of the modal responses of current data $C_j[i]$ and the baseline modal responses, $R_j[i]$, for each sensor are evaluated. $\gamma_j[i]$ vector is sorted in the increasing order for each sensor. The modal time-history responses are selected based on a user-defined threshold (i.e., <0.9). The sum of correlation coefficients, DL, with respect to each sensor is evaluated as follows:

$$DL[i] = \sum_{j=1}^{M} \gamma_j[i] \qquad (2.2)$$

The DL value of a sensor node falls low with the presence of damage in the vicinity of that node, as the correlation between the affected modal responses of the node with that of the baseline decreases. Therefore, the lowest DL value corresponds to the sensor node number nearest to the damage location and can be alarmed for further action. In the absence of damage, the DL values of all sensor nodes are nearly equal, and the current data become the new baseline data for the next arriving dataset for diagnosis. In case the current data are diagnosed as healthy, the "baseline data" are updated with the current data along with their corresponding cointegrating vectors. The flowchart of the proposed method is shown in Figure 2.6.

The damage is created at 1/3 span (approximately at element no. 5) by a static load through a hydraulic jack in small increments. The acceleration response of the beam due to minor damage is named as 'current data'. Few of the extracted sources of current data, after eliminating EOV, are shown in Figure 2.7. The normalized DL values for the raw signal and the pre-processed signal are indicated in Figure 2.8 (a) and (b) respectively. It can be seen from Figure 2.8 (b), that the location of the damage is shown by the reduced DL values, exactly near the 1/3 span length of the beam. From the results of the experimental study, it can be shown that the EOV is eliminated from the data using cointegration and the minor damage is successfully located using the damage index constructed from the modal responses extracted using improved SOBI technique.

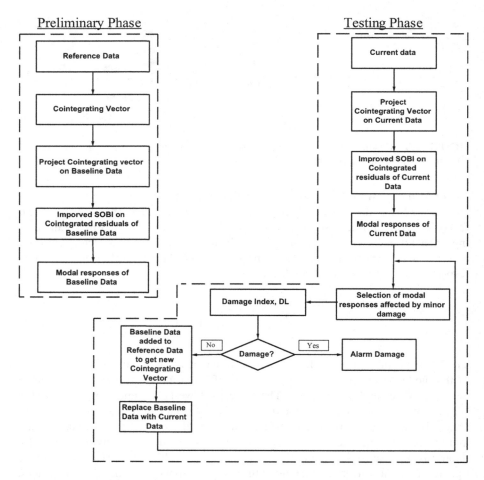

FIGURE 2.6 Flowchart of the proposed technique to handle EOV and online SHM.

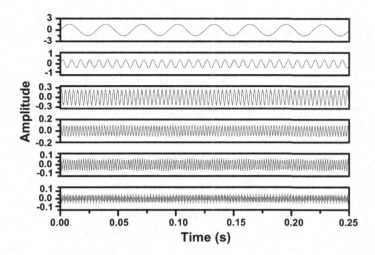

FIGURE 2.7   The sources extracted using blind source separation.

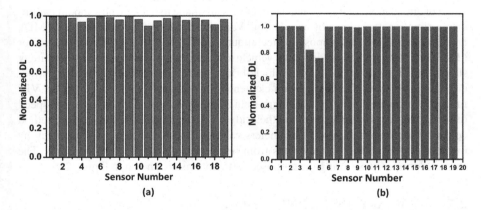

FIGURE 2.8   Normalized damage index: (a) using raw signal and (b) using preprocessed signal with proposed algorithm and cointegration.

2.4   MODAL IDENTIFICATION OF BRIDGES USING AN INSTRUMENTED MOVING VEHICLE IN DRIVE-BY HEALTH MONITORING

Vibration-based techniques have extensively been used for developing SHM techniques for bridges. Several vibration-based SHM techniques developed so far require instrumentation of the entire bridge with sensors and measuring the time-history responses. Instrumentation of the entire bridge is in fact laborious, expensive, and also very tedious to operate. As a cost-effective alternative to this traditional SHM, a technique is gaining popularity in the recent past where indirect measurements from an instrumented vehicle moving over the bridge are used for SHM [10]. SHM using these indirect measurements has several advantages over the more conventional direct approach with a fully instrumented bridge. Even though only a single sensor on a vehicle is used to collect the time-history data in the indirect approach, it offers much higher spatial resolution than the traditional direct approach. This is because the instrumented vehicle in the indirect approach can receive the vibration characteristics of each point as it passes over the bridge. Keeping these things in view, efforts are being made to develop novel modal identification and damage diagnostic techniques using indirect measurements.

The modal identification of a bridge through an instrumented test vehicle acting as a sensor is extremely challenging. The measured dynamic responses from the instrumented vehicle include components associated with the bridge and vehicle, as well as driving frequencies apart from the disturbances such as roughness of the bridge surface. Therefore, it is highly challenging to isolate these components and obtain the original bridge modal parameters through vehicle response. Since the vehicle frequency is not present in the contact-point response, it offers a clear measurement for deriving the modal frequencies and dynamic characteristics of the bridge. However, it is preferred to design the vehicle frequency higher than the interested bridge frequencies to handle the effect of attenuation. In the present work, efforts are made to devise a new modal parameter estimation technique by combining Variational Mode Decomposition (VMD), an advanced signal decomposition technique, with Teager Kaiser Energy Operator (TKEO). The numerical model of the vehicle as a quarter car is integrated with a simply supported bridge model to obtain a vehicle-bridge–interacted model. The bridge is modeled using the Euler–Bernoulli beam elements. The vibration responses from the VBI model are evaluated using the Newmark-Beta integration method for the specified speed of the quarter-car model, moving on the bridge. The outcomes of the proposed method confirm that the bridge modal parameters can be identified with a good accuracy by extracting bridge-related components from instrumented vehicle body responses. The extracted mode shapes using the proposed formulations are shown in Figure 2.9.

2.5 SUMMARY AND CONCLUSION

In this chapter, few of the recent advancements in vibration-based SHM of civil infrastructure are presented. Three signal decomposition techniques for multi-model approaches for civil structural health monitoring are presented for vibration-based SHM, which are shown to detect subtle damage. A novel technique suitable for online SHM handling EOV is also presented. Experimental investigations to demonstrate the effectiveness of the proposed techniques are presented briefly. Studies clearly indicate the effectiveness of the proposed techniques and their practical applicability.

Also, the recent advancements in indirect health monitoring using instrumented vehicle are introduced, and a signal processing technique for modal identification is presented. Mode shape extraction is performed using VMD and TKEO of the contact-point responses between the vehicle and the bridge. The results of the proposed method affirm that the bridge modal parameters can be identified with a good accuracy by extracting bridge-related components from instrumented vehicle body responses.

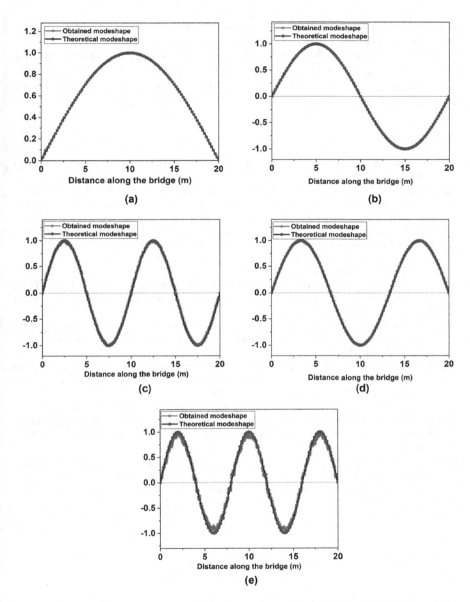

FIGURE 2.9 Mode shapes extracted from the proposed VMD-TKEO formulation using VBD. (c) should come first, and then (d)

ACKNOWLEDGMENTS

The author sincerely acknowledges Dr. A. Rama Mohan Rao, Rtd. Chief Scientist, CSIR-SERC, for his technical guidance and support in the research of multi-model techniques during his tenure in CSIR-SERC. Also, the financial support by Department of Science and Technology-SERB, Government of India, under POWER research grant SERB/F/130/2021–2022, for carrying out the research in drive-by health monitoring is deeply acknowledged. The author fondly acknowledges the project students, the project assistants, and the technical staff of SHML, CSIR-SERC, involved in this part of research and experimental activities. The chapter is also published with the permission of Director, CSIR-SERC.

REFERENCES

[1] Doebling, S.W., Farrar, C., and Prime, M.B., A summary review of vibration-based damage identification methods, *The Shock and Vibration Digest*, 30(2): 91–105, 1998.

[2] Lakshmi, K., and Rao, A.R.M., Damage identification technique based on time series models for LANL and ASCE benchmark structures, *Insight-Non-Destructive Testing and Condition Monitoring*, 57(10): 580–588, 2015.

[3] Lakshmi, K. and Rao, A.R.M., Structural damage detection using ARMAX time series models and cepstral distances, *Sādhanā*, 41(9): 1081–1097, 2016.

[4] Golyandina, N., and Zhigljavsky, A., Singular spectrum analysis for time series. *Springer Briefs in Statistics*. Springer; 2013.

[5] Lakshmi, K., Rao, A.R.M., and Gopalakrishnan, N., Singular spectrum analysis combined with ARMAX model for structural damage detection, *Structural Control and Health Monitoring*, 24(9): e1960, 2017.

[6] Belouchrani, A., Abed-Meraim, K., Cardoso, J., and Moulines, E., A blind source separation technique using second-order statistics, *IEEE Transactions on Signal Processing*, 45(2): 434–444, 1997.

[7] Krishnasamy, L., and Arumulla, R.M.R., Baseline-free hybrid diagnostic technique for detection of minor incipient damage in the structure, *Journal of Performance of Constructed Facilities*, 33(2): 04019018, 2019.

[8] Huang, N.E., *Hilbert-Huang Transform and its Applications*. World Scientific; 2014.

[9] Lakshmi, K., and Arumulla, R.M.R., A hybrid structural health monitoring technique for detection of subtle structural damage, *Smart Structures and Systems*, 22(5): 587–609, 2018.

[10] Cerda, F., Garrett, J., Bielak, J., Bhagavatula, R. and Kovačević, J., Exploring Indirect Vehicle-Bridge interaction for SHM. Proceedings of the Fifth International Conference on Bridge Maintenance, Safety and Management, IABMAS2010, Philadelphia, USA, 696–702; 2010.

3 3D Vibration Control of Flexible Manipulator Using Inverse System and Strain Feedback

Minoru Sasaki and Daiki Maeno
Gifu University

Muguro Joseph
Gifu University
Dedan Kimathi University of Technology

Mizuki Takeda
Gifu University

Waweru Njeri
Dedan Kimathi University of Technology

Kojiro Matsushita
Gifu University

3.1 INTRODUCTION

In recent years, industrial robots are required to be "lightweight" for energy saving and cost reduction, and "high speed" to improve work efficiency. However, since weight reduction leads to low rigidity, mechanical vibration is likely to occur, and if you aim for high-speed operation, vibration will occur more easily, so it is essential to improve control performance. For such flexible manipulators, conventional research has mainly focused on his two-dimensional motion vibration suppression research [1,2]. On the other hand, research on vibration suppression of three-dimensional motion has hardly been done due to the difficulty of modeling. One of the few studies on vibration control in three-dimensional space is the proposal of a control method using an inverse system [3–5], which shows that vibration can be suppressed, but the resonance frequency component in the torsional direction remains. There are still problems of stowing and overshooting against the target angle [6–10].

DOI: 10.1201/9781003402695-3

Therefore, in this research, we focus on the three-dimensional motion control of the flexible manipulator and aim to improve the vibration suppression performance in the torsional direction and the overshoot. First, we propose a stable inverse system using inner–outer decomposition for the unstable zeros of the flexible manipulator and try to verify its performance. Second, in order to remove the vibration components remaining after the application of the proposed inverse system, two-degree-of-freedom control that combines the inverse system and strain feedback control is used to improve vibration suppression while maintaining the ability to follow the target angle. Third, we show the effectiveness of the proposed method by applying it to tip trajectory control, which is often performed by ordinary industrial arms.

3.2   CONTROL SETUP

Figure 3.1 shows the flexible manipulator to be controlled. From Figure 3.1a, Joint1 rotates in the twisting direction, and Joint2 and Joint3 rotate in the bending direction. A strain gauge is attached to the base of each link. In addition, the trajectory of the arm tip is measured using optical motion capture OptiTrack™. Figure 3.1b shows the created 3D model using Maplesim™. Further details of the measurement and modeling are available in Ref. [3].

3.2.1   MODEL EVALUATION

This section provides a verification of the validity of the models. The evaluation targets are the nonlinear model and the linear model described so far, and the evaluation is performed by comparing with the actual measurement data obtained in the actual manipulator.

Figure 3.2 shows the strain response when the target angle of each motor is operated as a step response of $20°$. The steady-state strains are the same in the in-plane

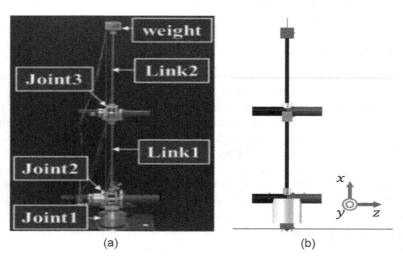

(a)   (b)

FIGURE 3.1   Flexible manipulator model.

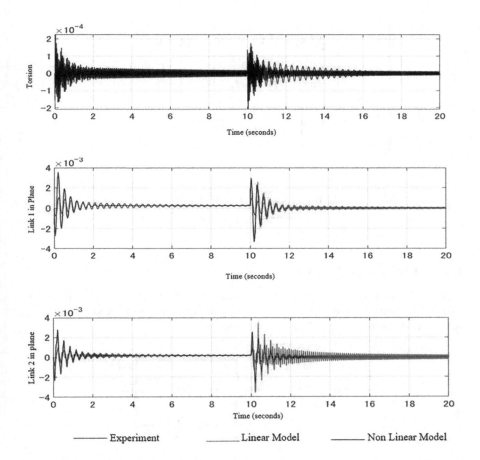

FIGURE 3.2 Comparison of strain response.

direction of links 1 and 2. It can be seen that the vibration characteristics also match. From the above, it can be evaluated that the model created this time is appropriate in terms of angular response and vibration characteristics. The controller will be designed based on the linear model created.

3.3 CONTROL SYSTEM DESIGN

3.3.1 INVERSE SYSTEM DESIGN

The input is the target angle of each motor, and the output is the state equation of the angular displacement and strain of the motor. Since the state equation has an unstable zero near the imaginary axis and the number of input and output variables is different, the inner–outer decomposition is performed by the extended Riccati equation (3.1) as discussed in Ref. [6,7]. Creating the inverse system of equation (3.2) from the outer function results in a stable system.

$$\begin{cases} X^T A_a + A_a^T X - X B_a B_a^T X + Q_a = 0 \\ Q_a = \begin{bmatrix} C^T C & C^T D \\ D^T C & D^T D - I_p \end{bmatrix}, E_a = \begin{bmatrix} I_n & 0 \\ 0 & 0 \end{bmatrix} \\ A_a = \begin{bmatrix} A & B \\ 0 & I_p \end{bmatrix}, B_a = \begin{bmatrix} 0 \\ -I_p \end{bmatrix} \end{cases} \tag{3.1}$$

$$\begin{cases} \dot{x}(t) = \left(A - BD^{-1}C\right)x(t) + BD^{-1}u(t) \\ y(t) = -D^{-1}Cx(t) + D^{-1}u(t) \end{cases} \tag{3.2}$$

3.3.2 Controller Design

Figure 3.3 shows a block diagram of the entire control system. The vibration of the resonance frequency component is reduced by the inverse system, and the remaining vibration component is removed by strain feedback.

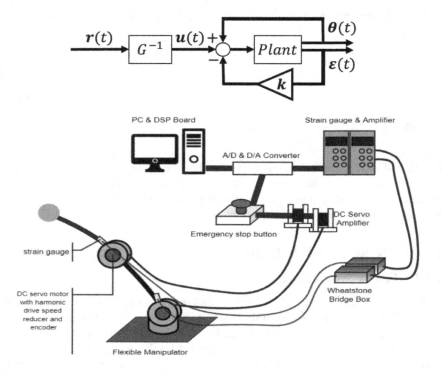

FIGURE 3.3   Block diagram of the control system and illustration of experimental setup.

3.4   EXPERIMENT RESULTS

Figure 3.4 shows the angular response when the target angle of each joint is input as a step response of 20°, and Figure 3.5 shows the strain value. It can be confirmed that the torsion and in-plane vibration at the start of motion are reduced compared to the conventional research. Furthermore, it can be seen that the vibration suppression performance is improved by adopting two-degree-of-freedom control.

Next, Figure 3.6 shows the trajectory when a circle is drawn at a speed of 1.25 rad/s as trajectory control, and control performance is compared with (1) inverse kinematics, (2) inverse system, and (3) two-degree-of-freedom control that combines inverse system and strain feedback. It can be confirmed that the trajectory following performance is maintained in (2) and (3) compared to (1). In addition, it can be seen that the vibration suppression performance is improved by adopting two-degree-of-freedom control.

Regarding the vibration suppression performance seen in Figure 3.7, it can be seen that the resonance frequency component of 3 Hz is suppressed by the inverse system, and the remaining vibration component is suppressed by the direct strain feedback controller (DSFB). From this, it was confirmed that the two-degree-of-freedom control by the inverse system and DSFB is also effective in suppressing tip vibration.

3.5   CONCLUSION

In this research, by applying a controller that combines a stable inverse system and strain feedback to a flexible manipulator, it was confirmed that performance for vibration suppression and angular response was improved compared to previous research. Then, we performed trajectory control and demonstrated that the proposed method is also effective for 3D vibration control at the tip of the arm.

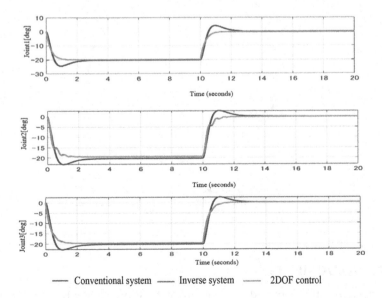

FIGURE 3.4   Angle response.

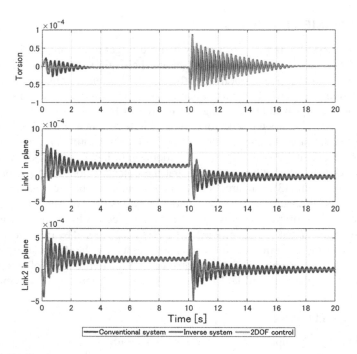

FIGURE 3.5   Strain response.

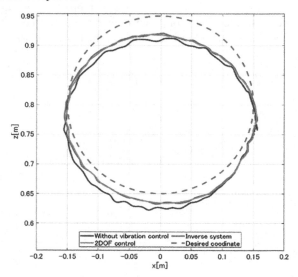

FIGURE 3.6   Tip coordinate.

ACKNOWLEDGMENTS

This work is partially supported by grants-in-aid for Promotion of Regional Industry–University–Government Collaboration from Cabinet Office, Japan.

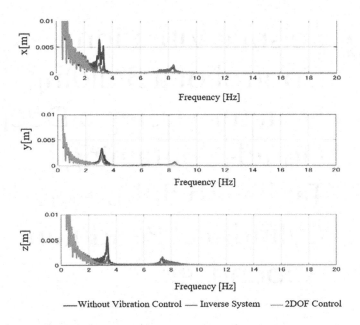

FIGURE 3.7 FFT of tip coordinate.

REFERENCES

[1] Njeri, W., Sasaki, M., and Matsushita, K., Enhanced vibration control of a multilink flexible manipulator using filtered inverse controller, *Robomech J*, Vol. 5, pp. 28 (2018).

[2] Njeri, W., Sasaki, M., and Matsushita, K., Two-degree-of-freedom control of a multilink flexible manipulator using filtered inverse feedforward controller and strain feedback controller, *2018 IEEE International Conference on Applied System Invention (ICASI)*, 2018, pp. 972–975.

[3] Njeri, W., Sasaki, M., and Matsushita, K., Gain tuning for high speed vibration control of a multilink flexible manipulator using artificial neural network, *Transaction of ASME Journal of Vibration and Acoustics*, Vol. 141, No. 4, pp. 11 (2019). doi:10.1115/1.4043241

[4] Kaneko, M., Special feature on flexible arms, *Journal of the Robotics Society of Japan*, Vol. 6, No. 5, pp. 414–466 (1988).

[5] Matsuno, F., Special issue on "flexible manipulators", *Journal of the Robotics Society of Japan*, Vol. 12, No. 2, pp. 169–230 (1994).

[6] Sasaki, M., Shimasaki, H., Itou, S., and Shimizu, T., Modeling and motion control of a flexible manipulator using a formula manipulation, *Proceedings of JSME Annual Conference on Robotics and Mechatronics*, Vol. 1A1–E18 (2010). doi: 10.1299/jsmermd.2010._1A1-E18_1

[7] Shimizu, T., Sasaki, M., and Okada, T., Hand position control of 2-DOF planar flexible manipulator based on extended dynamics, *Transactions of the Society of Instrument and Control Engineers*, Vol. 44, No. 5, pp. 389–395 (2008).

[8] Shimizu, T., and Sasaki, M., A study on passivity of 2-DOF flexible manipulator, *Transactions of the Society of Instrument and Control Engineers*, Vol. 38, No. 3, pp. 862–867 (2002).

[9] Lochana, K., Roy, B.K., and Subudhib, B., A review on two-link flexible manipulator, *Annual Reviews in Control*, Vol. 42, pp. 346–367 (2016).

[10] Sasaki, M., Asai, A., Shimizu, T., and Ito, S., Self-tuning control of a two-link flexible manipulator using neural networks, *Proceedings of ICCAS-SICE International Conference 2009*, pp. 2468–2473, 2009.

4 Seismic Vibration Control of a Building Structure Using a Deep Liquid-Containing Tank with Submerged Cylindrical Pendulum Appendage

Tanmoy Konar and Aparna (Dey) Ghosh
IIEST

4.1 INTRODUCTION

Passive tuned liquid dampers (TLDs) offer a cost-effective and reliable vibration control strategy for structures and have been implemented in some towers and sky-scrapers [1–4]. TLDs in the tank configuration employ the phenomenon of sloshing to dissipate energy and require a shallow liquid level in the tank for high volumetric efficiency. Here, it is pertinent to mention that a liquid-containing tank is termed a shallow tank when liquid depth in the tank is less than 0.1 times the length of the tank. For a deep tank, this ratio remains above 0.5. A high fraction of the residing liquid in a deep tank is impulsive liquid which synchronizes with the vibration of the tank and does not slosh. Further, the standing waveform observed in a vibrating deep tank is inefficient relative to the traveling waveform produced in a shallow tank [5]. That is why deep tanks are normally considered unsuitable for utilization as TLDs.

Most attempts to improve the effectiveness of deep tanks as TLDs include the utilization of flow-dampening devices, which can derive added damping through interference with the sloshing liquid [6]. This approach has a limitation as, in a deep tank, only a limited part of the total liquid sloshes. Recently, Konar and Ghosh [7] have set forth the concept of a deep liquid tank with submerged cylindrical pendulum appendage (DLT-CPA) to involve the impulsive liquid in the damping mechanism and thereby effectively utilize the deep tank as a passive damper. This is significant, as a large fraction of the total liquid in a deep tank behaves as impulsive liquid [8]. In a DLT-CPA, the cylindrical pendulum is kept submerged in the impulsive liquid zone.

DOI: 10.1201/9781003402695-4

When subjected to base excitation, the pendulum appendage undergoes oscillations and thereby energizes the dead or impulsive mass of the liquid. As a result, a fraction of the impulsive liquid mass behaves as an added mass to the pendulum appendage. Here, the damping mechanism is provided by the liquid drag on the pendulum as it oscillates under excitation.

In this chapter, first, the physical phenomenon and mathematical modeling related to the DLT-CPA are described. This is followed by the design of a DLT-CPA system for a seven-story example building. For the purpose of the damper design, the building is modeled as a single-degree-of-freedom (SDOF) structure having mass and natural period equal to the modal mass and period of the fundamental mode of the example building, which is its dominant mode. Thereafter, a numerical study is carried out to assess the performance of the DLT-CPA system by subjecting the SDOF structural system to recorded earthquake ground motions. The conclusions are drawn at the end.

4.2   DESCRIPTION OF DLT-CPA

A deep tank with a rectangular cross-section, as shown in Figure 4.1a and b, is considered. The liquid in the tank is water. A solid metal cylinder of uniform mass and cross-section is vertically suspended within the impulsive liquid zone from a hanging arrangement rigidly attached to the tank wall (see Figure 4.1b). The cylindrical appendage is fixed in such a way that it can oscillate like a pendulum when laterally excited. The oscillation period of the cylindrical appendage is tuned to the fundamental period of the host building. When the building is subjected to base excitation, a part of the excitation energy is transferred to the cylindrical pendulum appendage (CPA) through tuning. As a result, the cylindrical appendage is set into oscillating motion, which is opposed by the adjacent water through drag resistance, and this leads to energy dissipation. During the oscillation of the cylindrical appendage, a certain part of the impulsive liquid effectively functions as an added mass to the cylindrical appendage. Thus, in the

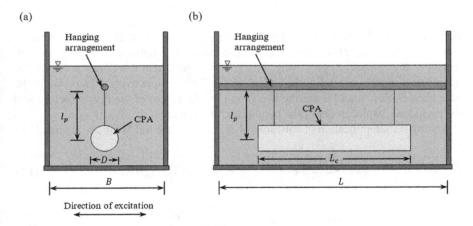

FIGURE 4.1   DLT-CPA: (a) front elevation and (b) side elevation.

DLT-CPA, the impulsive liquid participates in the damping mechanism. Further, while the water level in the upper part of the deep tank fluctuates, the performance of the proposed damper would not be affected so long as the cylinder appendage remains within the zone of the tank occupied by the impulsive liquid mass.

An analytical model of the DLT-CPA is developed considering the host structure as an SDOF system and with the following assumptions:

- The tank is rigid, and the liquid is incompressible.
- The cylindrical pendulum would be fully submerged in the liquid during oscillation.
- The appendage has all its mass concentrated at its center of mass.

The expression for the fundamental period of the submerged cylinder pendulum appendage, T_P, is given by [7]

$$T_P = 2\pi \sqrt{\frac{l_p (R+k)}{g (R-1)}} \tag{4.1}$$

Here, l_p, g, and k are the length of the pendulum, given by the distance between the hanging arrangement and the center of the pendulum, gravitational acceleration, and coefficient of added mass, respectively. R is given by ρ_P / ρ_l, where ρ_P and ρ_l are the density of the CPA and the damper liquid, respectively.

Now, for the derivation of the equations of motion of the SDOF-structure-DLT-CPA system, the dynamic equilibrium of the CPA and the mass of the SDOF system are considered. The equations obtained thereby are as follows [7]:

$$m_s \ddot{u} + \left(\rho_P + k \; \rho_l\right) L_c \frac{\pi D^2}{4} \left(\ddot{u} + l_p \; \ddot{\theta} \cos\theta\right) + k_s \; u + c_s \; \dot{u} = -m_s \ddot{Z} \tag{4.2}$$

$$I\left(\ddot{\theta} + \frac{\ddot{u} + \ddot{Z}}{l_p} \cos\theta\right) + l_p F_d + l_p F_g \sin\theta - l_p F_B \sin\theta = 0 \tag{4.3}$$

Here, m_s, c_s, and k_s are the mass, damping coefficient, and stiffness of the host SDOF system, respectively. D, L_c, and I denote the diameter, length, and mass moment of inertia of the cylinder, respectively. F_d, F_g, and F_B represent the drag force, gravitational force, and buoyancy force action on the CPA, respectively. $\ddot{Z}$ represents the base acceleration. u and θ are the horizontal displacement of the host structure and the rotational displacement of the pendulum, respectively.

On further simplification of equations (4.2) and (4.3), the following are obtained [7]:

$$\ddot{u} + \frac{2\zeta_s \omega_s}{(1+\mu\lambda)} \dot{u} + \frac{\omega_s^2}{(1+\mu\lambda)} u = -\ddot{Z} - \frac{\mu\lambda}{(1+\mu\lambda)} l_p \ddot{\theta} \cos\theta \tag{4.4}$$

$$\ddot{\theta} + \frac{2l_p}{\pi D} \frac{C_d}{(R_\rho + k)} |\dot{\theta}| \dot{\theta} + \omega_p^2 \sin\theta = -\frac{(\ddot{u} + \ddot{Z})}{l_p} \cos\theta \tag{4.5}$$

Here, ω_s, ζ_s, ω_p, μ, and C_d denote the structural frequency, the structural damping ratio, the frequency of the pendulum, the mass ratio, and the drag coefficient, respectively. λ is given by equation (4.6). Here, it is pertinent to mention that C_d and k chiefly depend on the ratio L_c / D.

$$\lambda = \left(1 + \frac{k}{R}\right) \tag{4.6}$$

4.3 DESCRIPTION OF EXAMPLE STRUCTURE

The numerical study is carried out on a seven-story dormitory building. The plan of a typical floor of the example building is depicted in Figure 4.2. The building consists of a reinforced concrete frame with walls made of aerated concert blocks. The story height of the building is 3.6 m. The distance between the foundation level and the plinth level of the building is 2.5 m. The cross-sections of columns and beams are considered 0.375 m × 0.375 m and 0.250 m × 0.400 m, respectively. The slabs are assumed to be 0.125 m thick. The external and internal walls are 0.250 and 0.100 m thick, respectively. The doors and windows are assumed to cover 30% of the total area of the walls. The grades of concrete and reinforcement steel are assumed as Fe500 and M30, respectively. The densities of aerated concert block and concrete are taken as 6.5 and 25 kN/m³, respectively. As per IS 875 (Part 2)-1987, the live load on the corridors, passages, staircases, and room for indoor sports of a dormitory building is 3 kN/m³, whereas the same on the bedrooms, toilets, and baths is 2 kN/m³ [9]. In this work, for simplicity, a constant value of the live load equal to 2.5 kN/m³ is taken on the entire floor. The structural damping ratio, ζ_s, is considered as 2%. Here, it is pertinent to mention that the recommended damping ratio for reinforced concrete structure as per IS 1893 (Part 1)-2016 is 5% [10]. This recommendation is based on the consideration that the structure would experience inelastic deformations under seismic excitation

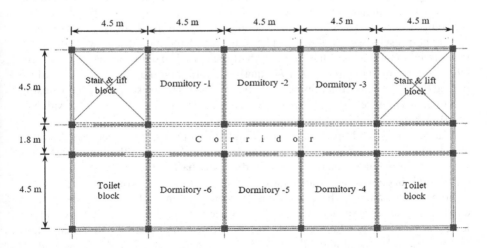

FIGURE 4.2 Typical floor plan of the example dormitory building.

which represents the limit state of collapse [10]. However, in the present work, the damper is used to reduce structural displacement and thereby improve serviceability. Here, the deformation of the structure remains within the elastic limit. In such case, the intrinsic damping of the structure is much low. Some design codes specifically recommend a lower value of damping ratio for the serviceability limit state. For example, the Australian/New Zealand Standard recommends a 1.5% damping ratio for concrete structures for the serviceability limit state [11]. Thus, the consideration of 2% damping ratio for the present work is justifiable.

The building is modeled as a three-dimensional moment-resisting frame in STAAD.Pro V8i SS6 graphical environment [12]. The fundamental period of the building is determined as 1.983 s ($\omega_s = 3.167$ rad/s), and the effective mass participating in the fundamental mode is equal to 1221970 kg. Here, it is pertinent to mention that the modal participation factor of the fundamental mode is 83.08%. For the determination of the responses of the structure-DLT-CPA system and the structure alone under seismic base excitations, the building is considered as an equivalent SDOF system with natural period and mass, respectively, equal to the fundamental time period and the effective mass participating in the fundamental mode of the example building [13].

4.4 DESIGN OF DEEP LIQUID-CONTAINING TANK WITH CPA

For the design of the DLT-CPA system, a practically feasible value of the mass ratio, μ, equal to 0.6% is considered. For such a low μ, the tuning ratio, γ, normally has a value close to unity. Thus, for simplicity, $\gamma = 1$ is assumed for the present design. Hence, the oscillation period of the cylindrical appendage is the same as the fundamental period of the equivalent SDOF structure.

Let us assume that the DLT-CPA system comprises six identical units. Thus to provide the total considered mass ratio of 0.6%, the mass of each cylindrical appendage would be 0.1% of the mass of the SDOF system. It is assumed that the cylindrical appendage is made of copper, having a density equal to 8960 kg/m³. The liquid in the tank is water with a density equal to 997 kg/m³. Thus, the density ratio, R, for the present design is 8.987. The diameter of the cylindrical appendage, D, is assumed to be 0.25 m. Thus, the length of the cylindrical appendage, L_c, is calculated as 2.778 m. This makes L_c / D equal to 11.113, and the corresponding values of k and C_d are determined as 0.975 [14] and 0.826 [15], respectively. Now, using equation (4.1), the length of the pendulum, l_p, is calculated as 0.784 m. The value of g is 9.81 m/s².

The length, L, and breath, B, of the tanks of the DLT-CPA system are assumed as 3.5 and 2 m, respectively. The maximum depth of water in the tank is taken as 2.5 m. The appendage has to be fixed in the lower half of the tank, i.e., in the zone of the tank occupied by impulsive liquid mass. Further, to avoid the effect of the bottom of the tank on the oscillation frequency of the submerged pendulum, a clearance of 1.5 D is to be provided between the bottom surface of the tank and the lowest point of the pendulum [7,14]. To satisfy this condition, the hanging arrangement of the cylindrical appendages is fixed at a height of 1.3 m above the bottom surface of the tank.

4.5 PERFORMANCE ASSESSMENT OF DLT-CPA

The potential of the designed DLT-CPA system as a damping device is assessed by subjecting the equivalent SDOF structural system to recorded earthquake ground motion. For this purpose, three recorded ground motions, namely the 90° component of the 1989 Loma Prieta earthquake (recording station: Capitola) with a peak ground acceleration (PGA) of 0.439 g, the 291° component of the 1971 San Fernando earthquake (recording station: Castaic – Old Ridge Route) with a PGA of 0.275 g, and the 225° component of the 1979 Imperial Valley earthquake (recording station: Calexico Fire Station) with a PGA of 0.277 g, are considered [16]. The choice of the recorded ground motions is made in such a way that the fundamental period of the example building falls within the range of the dominant period of the excitations.

The equations of motion of the structure-DLT-CPA system given by equations (4.4) and (4.5) are solved numerically for the considered base excitations using MATLAB with the help of its inbuilt 'ode45' solver [17]. The effectiveness of the damper is assessed by the achieved reduction in the displacement response of the equivalent SDOF structural system expressed in percentage with respect to the uncontrolled response.

4.6 RESULT AND DISCUSSION

The reductions by the DLT-CPA system in root-mean-square (rms) and peak displacement of the equivalent SDOF system for the considered seismic excitations are summarized in Table 4.1. It can be noted from Table 4.1 that the DLT-CPA system is able to obtain reductions of 11.3%–14% and 21.8%–36.2%, respectively, in the peak and rms displacement response of the equivalent SDOF structural system. This is significant for a passive control system with a mass ratio of only 0.6%. An indicative set of time histories of displacement of the controlled and the uncontrolled structural system under the Loma Prieta earthquake excitations is shown in Figure 4.3. The rotational displacement of the submerged pendulum appendage of the DLT-CPA system is also studied, and its peak values under the considered ground motions are 31.2°–44.5°. This indicates significant interaction between the pendulum appendage and the liquid in the deep tank. A time history of the rotational displacement of the

---

TABLE 4.1

Structural Displacement Response Reduction by the DLT-CPA System

Earthquake	Displacement of Uncontrolled Structure		Reduction by DLT-CPA System (%)		Peak Rotational Displacement of the Submerged Pendulum Appendage (°)
	rms	Peak	rms	Peak	
Loma Prieta	0.045	0.132	25.9	14.0	44.5
San Fernando	0.029	0.082	36.2	13.1	31.2
Imperial Valley	0.036	0.081	21.8	11.3	32.2

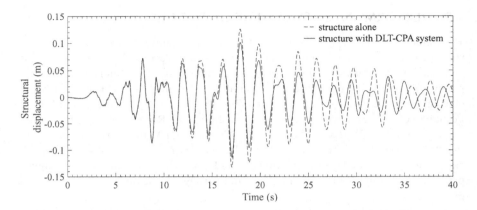

FIGURE 4.3 Time histories of displacement of the controlled and the uncontrolled structural system under the Loma Prieta earthquake.

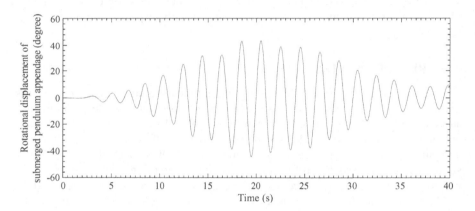

FIGURE 4.4 Time history of rotational displacement of submerged pendulum appendage of DLT-CPA system under the Loma Prieta earthquake.

submerged pendulum appendage under the Loma Prieta earthquake excitations is indicated in Figure 4.4.

Here it is pertinent to mention that, like other inertia-based dampers, the performance of the DLT-CPA would improve with a higher mass ratio. However, in the case of building structures, normally μ is kept in the range of 0.5%–1.5% due to restrictions in addition to gravity load on the structure and space constraint [3]. In view of this, the value of the mass ratio is taken as 0.6% in the present study.

4.7 CONCLUSIONS

Deep liquid-containing tanks, which are generally considered unsuitable for use as TLDs, can be effectively utilized for seismic protection of buildings by installing a CPA, submerged within the tank, and tuning the oscillating period of the appendage to the dominant period of the host building. The performance of the proposed system is not influenced by the variation in water depth within the tank due to filling and distribution of water, so long the appendage remains within the impulsive zone of liquid. This provides the option of using functional deep tanks, such as overhead water tanks as simple passive vibration control systems. The necessary modifications do not hamper the tank's functionality. The reductions by the proposed system in rms and peak displacement response of the equivalent SDOF system are significant for a passive supplemental damping system.

ACKNOWLEDGMENTS

This work is not funded by any funding agency, and there is no conflict of interest.

REFERENCES

[1] Tamura Y, Fujii K, Ohtsuki T, Wakahara T, Kohsaka R. Effectiveness of tuned liquid dampers under wind excitations. *Eng Struct* 1995;17:609–21.

[2] Soto MG, Adeli H. Tuned mass dampers. *Arch Comput Methods Eng* 2013;19:419–31. https://doi.org/10.1007/978-3-319-72541-3_18.

[3] Konar T, Ghosh AD. A review on various configurations of the passive tuned liquid damper. *J Vib Control* 2022;29(9–10):1945–1980. https://doi.org/10.1177/10775463221074077.

[4] Konar T, Ghosh AD. A study on the economic perspective of utilizing liquid tanks as dynamic vibration absorbers in building structures. In: Babu SA, editor. *5th World Congress Disaster Management.* Vol. I, London, UK: Routledge; 2022. https://doi.org/10.4324/9781003341956.

[5] Marsh A, Prakash M, Semercigil E, Turan ÖF. A numerical investigation of energy dissipation with a shallow depth sloshing absorber. *Appl Math Model* 2010;34:2941–57. https://doi.org/10.1016/j.apm.2010.01.004.

[6] Konar T, Ghosh AD. Flow damping devices in tuned liquid damper for structural vibration control: a review. *Arch Comput Methods Eng* 2021;28:2195–207. https://doi.org/10.1007/s11831-020-09450-0.

[7] Konar T, Ghosh AD. Adaptation of a deep liquid-containing tank into an effective structural vibration control device by a submerged cylindrical pendulum appendage. *Int J Struct Stab Dyn* 2021;21:2150078. https://doi.org/10.1142/S0219455421500784.

[8] *ACI 350/350R-01 Environmental Engineering Concrete Structures and Commentary.* Farmington Hills, Michigan, USA: 2001.

[9] *IS 875 (Part 2) Code of Practice for Design Loads (Other than Earthquake) for Buildings and Structures - Imposed Loads.* New Delhi: Bureau of Indian Standards; 1987.

[10] *IS 1893 (Part 1) - Criteria for Earthquake Resistant Design of Structures : General Provisions and Buildings.* New Delhi: Bureau of Indian Standards; 2016.

[11] *AS- NZS 1170-2: Structural Design Actions - Part 2: Wind Actions.* Vol. 2. Standards Australia Limited/Standards New Zealand; 2011.

[12] *STAAD Pro. V8i SS6.* Bentley Systems; 2015.

[13] Konar T, Ghosh AD. Design of overhead water tank with floating base for utilization as tuned liquid damper against lateral excitation. In: Maiti DK, Jana P, Mistry CS, Ghoshal R, Afzal MS, Patra PK, et al., editors. *Recent Advances in Computational and Experimental Mechanics (Select Proc. ICRACEM 2020),* Vol. II, Singapore: Springer; 2022, pp. 665–72. https://doi.org/10.1007/978-981-16-6490-8_54.

[14] Kaneko S, Nakamura T, Inada F, Kato M, Ishihara K, Nishihara T, et al. *Flow-Induced Vibrations - Classifications and Lessons from Practical Experiences* (Second Edition). London, UK: Academic Press; 2014. https://doi.org/10.1016/b978-0-08-098347-9.00008-4.

[15] Blevins RD. *Applied Fluid Dynamics Handbook.* New York: Van Nostrand Reinhold Co; 1984.

[16] PEER. PEER Ground Motion Database -Pacific Earthquake Engineering Research Center; n.d. https://ngawest2.berkeley.edu/.

[17] MATLAB R2015a. 2015.

5 Seismic Analysis of Base-Isolated Liquid Storage Tanks Using Supplemental Clutching Inerter

Ketan Narayanrao Bajad, Naqeeb Ul Islam, and Radhey Shyam Jangid
Indian Institute of Technology (IIT) Bombay

5.1 INTRODUCTION

Structural vibrations induced due to wind or earthquake loads can result in severe deformations, fatigue, user discomfort, or even the collapse of civil engineering structures. Various structural vibration control techniques (passive, active, semi-active, and hybrid) have been utilized to reduce structural vibrations brought on by human activity or natural causes. The force response of passive devices is controlled using a spring, dashpot, and mass mechanism without needing an external power source. In general, active, semi-active, and hybrid are controlled devices which require sensors and actuators to operate. The passive vibration control system is more widely used than the semi-active and active control systems, which were exclusively used in specific situations. Passive control devices reduce structural vibrations by adding a flexible layer between the foundation and superstructure (base isolator), regulating the structure's frequency (tuned mass damper), and dissipating the energy from the excitations (viscous damper) [1].

However, using passive devices to control vibration appears to have fundamental limitations, which prompted the need to improve existing devices and bring new innovative concepts for vibration control of structures. Several inerter-centered devices were created after Smith's 2002 invention of the two-terminal inerter element [2]. Inerter-based devices have grown in prominence, especially during the past five years, and have been crucial in improving the performance of traditional structural systems. There have been numerous inertial mass dampers (IMD) developed, proposed, and studied recently, including gyro-mass dampers (GMD) [3], rotational inertial viscous dampers (RIVD) [4], viscous inertial mass dampers (VIMD) [5], tuned viscous mass dampers (TVMD) [6], adaptive tuned viscous inerter dampers

DOI: 10.1201/9781003402695-5

(ATVID) [7], electromagnetic inertial mass dampers (EIMD) [8], clutching inerters (CI) [9], and spring-dashpot-inerter system (SDIS) [10].

The CI proposed by Makris and Kampas is a combination of IMD and a clutch which, under attached conditions, have the resistive force proportionate to the difference in the relative acceleration between the two terminals. When detached, the resisting force of the CI becomes zero. Such exclusive characteristics of the CI enable the researcher to employ it in the response reduction of linear SDOF structures [11], SDOF and MDOF steel structures [12], rocking structures [13], base-isolated (BI) building structures [14], and BI bridge structures [15]. However, research has yet to be presented to study the performance of inertial devices for BI liquid storage tanks. The base-isolation technology in storage tanks causes an increase in the sloshing height response as the isolation period gets closer to the period of the sloshing. As a result of the isolator's added instability, the storage tank may fail.

This study assessed and investigated the seismic response of a BI tank with supplemental CI subjected to three real earthquake excitations. The study aims to evaluate the seismic response quantities, viz. sloshing displacement, sloshing acceleration, bearing displacement, and total shear force of the BI tank with and without the supplemental CI.

5.2 MODELING OF ISOLATED TANK WITH THE CLUTCHING INERTER

5.2.1 Mathematical Model of Isolated Liquid Storage Tank

A BI liquid storage tank is demonstrated schematically in Figure 5.1. The base and foundation are installed with a laminated rubber isolator (LRB). The fluid inside the tank is assumed to be incompressible and inviscid, the tank's walls are rigid, and the displacements caused by the fluid are minimal in an idealized lumped-mass model of the tank [16]. When the entire tank is excited, the liquid vibrates in three forms: convective (sloshing), impulsive, and stiff mass. The letters m_c, m_i, and m_r represent the convective, impulsive, and rigid masses. The damping constant of the convective and impulsive masses is defined by c_c and c_i, respectively; k_c and k_i represent the equivalent spring having the stiffness of the convective and impulsive masses. The tank is three degrees of freedom lumped-mass problem in which $(d_c, d_i, \text{ and } d_b)$ stand for the absolute displacements of the convective, impulsive, and rigid masses, respectively, and are subject to unidirectional excitations. Compared to the actual weight of the liquid inside the tank, the tank's self-weight is insignificant. According to Shrimali and Jangid [17], the governing equation of motion is expressed as follows:

$$[m]\{\ddot{u}\}+[c]\{\dot{u}\}+[k]\{u\} = -[m]\{r\}\ddot{u}_g \tag{5.1}$$

where $\{u\} = \{u_c, u_i, u_b\}^T$ is the displacement vector; $u_c = d_c - d_b$ and $u_i = d_i - d_b$ are the displacement of the convective and impulsive masses, respectively; $u_b = d_i - u_g$ is the displacement of the rubber bearing relative to the ground; $\{r\} = \{\ 0\quad 0\quad 1\ \}^T$ is the influence coefficient vector; $\ddot{u}_g$ is the earthquake ground acceleration and

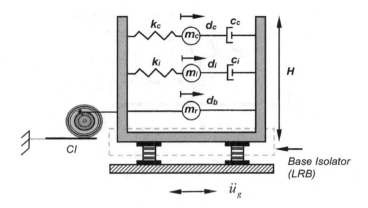

FIGURE 5.1   Schematic representation of the isolated tank with the CI.

T denotes the transpose of the matrix; $[m]$, $[c]$, and $[k]$ are the mass, damping, and stiffness matrices of the system, respectively, which are expressed as follows [17]:

$$[m] = \begin{bmatrix} m_c & 0 & m_c \\ 0 & m_i & m_i \\ m_c & m_i & M \end{bmatrix} = \begin{bmatrix} m_c & 0 & m_c \\ 0 & m_i & m_i \\ m_c & m_i & m_c + m_i + m_r \end{bmatrix} \quad (5.2)$$

$$c = diag\begin{bmatrix} c_c & c_i & c_b \end{bmatrix} = diag[\ 2\xi_c m_c \omega_c \quad 2\xi_i m_i \omega_i \quad c_b\ ] \quad (5.3)$$

$$k = diag\begin{bmatrix} k_c & k_i & k_b \end{bmatrix} = diag[\ m_c \omega_c^2 \quad m_i \omega_i^2 \quad m_b \omega_b^2\ ] \quad (5.4)$$

The restoring force of the LRB is $\{F_b\} = c_b \dot{u}_b + k_b u_b$.

5.2.2   MATHEMATICAL MODEL OF CLUTCHED INERTER

This work considers the interaction between an LRB and a supplemental linear CI. As depicted in Figure 5.2, the CI comprises a rachet, gear, and flywheel, with the rachet positioned between the gear and the flywheel. The rachet, however, rotates in the opposite direction from the flywheel. The rachet, which prevents the flywheel from pushing the structure, causes the CI's clutching effect. Contrary to conventional inerters, the CI can accelerate the rate at which velocity decays, reducing structures' displacement sensitivity. The CI is positioned at the isolator level and applies a resistive force simultaneously, but the LRB supplies stiffness and damping. The governing equation of motion of the BI tank with added CI is stated as follows [15]:

$$[m]\{\ddot{u}\} + [c]\{\dot{u}\} + [k]\{u\} + \{F_D\} = -[m]\{r\}\ddot{u}_g \quad (5.5)$$

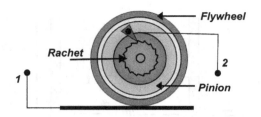

FIGURE 5.2   Schematic representation of the components of the CI.

Here, the resistive force of the CI is expressed as follows:

$$F_D = \begin{cases} b\ddot{u}_d & \text{when} \ \ \text{sgn}(\dot{u}_d\ddot{u}_d) > 0 \\ 0 & \text{when} \ \ \text{sgn}(\dot{u}_d\ddot{u}_d) \le 0 \end{cases} \tag{5.6}$$

where b is the constant of proportionality known as the CI's inertance, which is represented in the form of the total mass of the BI tank and is known as the inertance mass ratio of the CI $(\beta = b/M)$.

5.3   PROBLEM STATEMENT AND NUMERICAL STUDY

The aspect ratio $(S = H / R)$, the total height of the liquid stored in the BI tank (H), the radius of the tank (R), the damping ratios of the sloshing (ξ_c) and impulsive masses (ξ_i), the time period of the LRB (T_b), and the damping ratio of the LRB (ξ_b) are the parameters in the current study that determine the model of the BI tank. In this study, a broad tank is considered for a parametric investigation, and its characteristics are shown in Table 5.1. The tank wall is assumed to be made of steel, and the mass density and elastic modulus are calculated accordingly. In Table 5.2, the parameter values are tabulated. The outcomes are compared with those of Shrimali and Jangid for the isolated tank [17].

Three parameters, namely, isolation time period (T_b), damping ratio of the isolator (ξ_b), and inertance mass ratio, describe the isolation system and the CI considered for the tank. The values of the parameters are tabulated in Table 5.2. The parameters are expressed as follows:

$$T_b = 2\pi\sqrt{\frac{M}{k_b}} \ \text{and} \ \xi_b = \frac{C_b}{2M\omega_b} \tag{5.7}$$

Here $\omega_b = 2\pi / T_b$ is the frequency of the LRB and M is the effective mass of the liquid retained in the tank.

5.4   EARTHQUAKE CHARACTERISTICS

The seismic response of the isolated tank is investigated for Kobe (1995), Loma Prieta (1989), and Imperial Valley (1940). The characteristics of the earthquake excitations are tabulated in Table 5.3, and the time history of earthquake excitations is shown in Figure 5.3.

TABLE 5.1
Properties of the Cylindrical Tank [17]

Type of Tank	Material	H (m)	$S = H/R$	ω_c (Hz)	ω_i (Hz)	ξ_c (%)	ξ_i (%)	E (MPa)	ρ (kg/m^3)
Broad tank	Steel	14.6	0.6	0.148	5.757	0.5	2	200	7900

TABLE 5.2
Properties of the LRB and the CI [17]

Type of Isolator	Type of Inerter	T_b (s)	ξ_b (%)	β
Laminated rubber bearing	Clutching inerter	2	0.1	0.30

TABLE 5.3
Characteristics of the Earthquake Excitations

Earthquake Excitations	Recording Station	Duration (s)	Component	PGA (g)
Kobe, Japan (1995)	JMA	150	N00E	0.834
Loma Prieta, USA (1989)	Los Gatos Presentation Centre	25.005	N00E	0.570
Imperial Valley, USA (1940)	El Centro	53.76	S00E	0.348

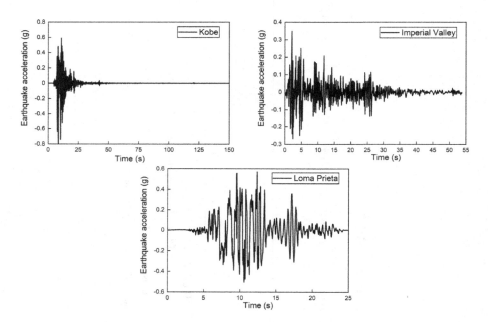

FIGURE 5.3 Time history of earthquake excitations considered in the study.

5.5 RESPONSE OF THE BASE-ISOLATED LIQUID STORAGE TANK

Bearing displacement, sloshing displacement, sloshing acceleration, and total base shear force are the response quantities estimated from the investigation. The responses of the uncontrolled isolated tank with and without the supplemental inerter system are evaluated to determine the efficiency of the CI. MATLAB code using the state-space method has been developed to determine the response quantities. First-order differential equations connect a set of input, output, and state variables in the state-space model of a physical system.

Figure 5.3 shows the time history of the response quantities in a broad, isolated tank subjected to the normal Kobe earthquake component. For the parameters shown in Tables 5.1 and 5.2, the response of the uncontrolled isolated tank with supplemental CI is compared. This indicates that the CI effectively minimizes the bearing and sloshing displacement by 44% and 26%. This is the clutching effect that the CI provides to the structure. Additionally, it is observed that the overall base shear response has been significantly reduced by nearly 31% when compared to base isolation with LRB. Thus, the lateral force acting on the BI tank is reduced by an identical amount.

Furthermore, Figure 5.4 shows that the base isolation with LRB is highly successful in lowering the tank's sloshing acceleration by 50%, proving the efficiency of LRB. The advantage CI offers in reduced bearing displacement and total base shear enhancing the design can offset the sloshing acceleration increase.

Similar trends for the four response quantities are seen when the identical tank is subjected to the Loma Prieta (1989) and Imperial Valley (1940) earthquake excitations. Figures 5.5 and 5.6 depict the same, respectively. Thus, combining the CI with an LRB isolation system can reduce the bearing displacement, sloshing displacement, and total base shear. This can be done without compromising the ability of the isolated system to reduce acceleration. When considering the isolation and inerter characteristics indicated in Tables 5.2 and 5.3, respectively, Table 5.4 illustrates the peak response values for the broad tank under various real-time earthquake excitations.

5.6 CONCLUSION

In this study, the response quantities subjected to three real earthquake excitations are calculated numerically using a BI tank model with the CI. Furthermore, the

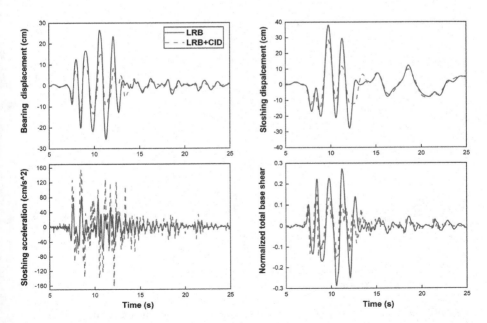

FIGURE 5.4 Time history of response quantities considered in the study subjected to Kobe, Japan (1995) earthquake excitation.

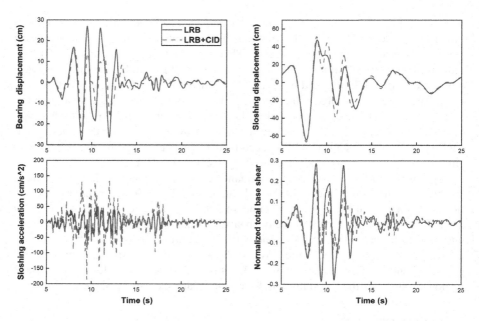

FIGURE 5.5 Time history of response quantities considered in the study subjected to Loma Prieta, USA (1989) earthquake excitation.

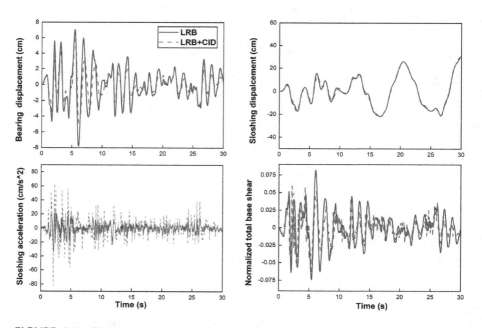

FIGURE 5.6 Time history of response quantities considered in the study subjected to Imperial Valley, USA (1940) earthquake excitation.

TABLE 5.4

Peak Earthquake Response of Isolated Broad Tank (LRB) with and without CI

Earthquake Excitation	Isolated Tank (LRB)	Isolated Tank with CI (LRB + CI)	Percentage Change	Observation
Bearing displacement response (cm)				
Kobe, Japan (1995)	26.55	14.89	44	Significant reduction
Loma Prieta, USA (1989)	27.55	24.33	12	in response
Imperial Valley, USA (1940)	7.80	6.39	18	
Sloshing displacement response (cm)				
Kobe, Japan (1995)	38.04	28.38	26	Significant reduction
Loma Prieta, USA (1989)	66.40	63.94	3.8	in response
Imperial Valley, USA (1940)	45.23	44.49	1.6	
Normalized total base shear (N)				
Kobe, Japan (1995)	0.29	0.20	31	Significant reduction
Loma Prieta, USA (1989)	0.29	0.22	24	in response
Imperial Valley, USA (1940)	0.08	0.06	25	

comparative study of various response quantities is carried out, with and without supplemental CI.

The current study leads to the following conclusions:

1. Peak bearing displacement and peak shear force significantly decrease when the CI is applied with the isolator, which establishes the effectiveness of the CI.
2. Although there is a minimal decrease sloshing displacement response of the tank, the issue of water spilling and roof tank damage may still be adequately controlled.
3. The hybrid (LRB + CI) system needs to address a little increase in the sloshing acceleration response, which can be accomplished by incorporating a different component or damper for sloshing acceleration recovery into the hybrid system.
4. The CI combines IMD and clutch, a better option than conventional inerter-based systems.

REFERENCES

[1] Madhekar S, Matsagar V. *Passive vibration control of structures*. Boca Raton: CRC Press; 2022. https://doi.org/10.1201/9781315269269.
[2] Ma R, Bi K, Hao H. Inerter-based structural vibration control: A state-of-the-art review. *Engineering Structures* 2021; 243:112655. https://doi.org/10.1016/j.engstruct.2021.112655.
[3] Saitoh M. On the performance of gyro-mass devices for displacement mitigation in base isolation systems. *Struct Control Health Monitoring* 2012; 19:246–59. https://doi.org/10.1002/stc.419.

[4] Hwang JS, Kim J, Kim YM. Rotational inertia dampers with toggle bracing for vibration control of a building structure. *Engineering Structures* 2007; 29:1201–8. https://doi.org/10.1016/j.engstruct.2006.08.005.

[5] Lu L, Duan YF, Spencer BF, Lu X, Zhou Y. Inertial mass damper for mitigating cable vibration. *Structural Control Health Monitoring* 2017; 24:e1986. https://doi.org/10.1002/stc.1986.

[6] Ikago K, Saito K, Inoue N. Seismic control of single-degree-of-freedom structure using a tuned viscous mass damper. *Earthquake Engineering and Structural Dynamics* 2012; 41:453–74. https://doi.org/10.1002/eqe.1138.

[7] Ali Sadeghian M, Yang J, Wang X, Wang F. Novel adaptive tuned viscous inertance damper (ATVID) with adjustable inertance and damping for structural vibration control. *Structures* 2021; 29:814–22. https://doi.org/10.1016/j.istruc.2020.11.050.

[8] Nakamura Y, Fukukita A, Tamura K, Yamazaki I, Matsuoka T, Hiramoto K, et al. Seismic response control using electromagnetic inertial mass dampers. *Earthquake Engineering and Structural Dynamics* 2014; 43:507–27. https://doi.org/10.1002/eqe.2355.

[9] Makris N, Kampas G. Seismic protection of structures with supplemental rotational inertia. *Journal of Engineering Mechanics* 2016;142:04016089 1–11. https://doi.org/10.1061/(ASCE)EM.1943-7889.0001152.

[10] Basili M, de Angelis M, Pietrosanti D. Modal analysis and dynamic response of two adjacent single-degree-of-freedom systems linked by spring-dashpot-inerter elements. *Engineering Structures* 2018;174:736–52. https://doi.org/10.1016/j.engstruct.2018.07.048.

[11] Wang M, Sun F. Displacement reduction effect and simplified evaluation method for SDOF systems using a clutching inerter damper. *Earthquake Engineering and Structural Dynamic* 2018;47:1651–72. https://doi.org/10.1002/eqe.3034.

[12] Málaga-Chuquitaype C, Menendez-Vicente C, Thiers-Moggia R. Experimental and numerical assessment of the seismic response of steel structures with clutched inerters. *Soil Dynamics and Earthquake Engineering* 2019;121:200–11. https://doi.org/10.1016/j.soildyn.2019.03.016.

[13] Thiers-Moggia R, Málaga-Chuquitaype C. Seismic protection of rocking structures with inerters. *Earthquake Engineering and Structural Dynamics* 2019;48:528–47. https://doi.org/10.1002/eqe.3147.

[14] Barkale S, Jangid RS. Performance of clutched inerter damper for base-isolated structures under near-fault motions. *Engineering Research Express* 2022;4:035016. https://doi.org/10.1088/2631-8695/ac8278.

[15] Jangid RS. Seismic performance assessment of clutching inerter damper for isolated bridges. *Practice Periodical on Structural Design and Construction* 2021;27(2):04021078. https://doi.org/10.1061/(ASCE)SC.1943.

[16] Housner, GW., and M. A. Haroun. Dynamic analyses of liquid storage tanks, In *Proceeding of The Seventh World Conference of Earthquake Engineering*, Istanbul, Turkey, 1980;8:431–438.

[17] Shrimali MK, Jangid RS. Seismic response of base-isolated liquid storage tanks. *Journal of Vibration and Control.* 2003;9(10):1201–18. https://doi:10.1177/107754603030612.

6 Optimal Piezo Patch Placement Using Genetic Algorithm and Performance Evaluation of Various Optimal Controllers for Active Vibration Control of Cantilever Beam

Yusuf Khan
Fr. Conceicao Rodrigues Institute of Technology
Anjuman-I-Islam's Kalsekar Technical Campus

S. M. Khot
Fr. Conceicao Rodrigues Institute of Technology

Khan Nafees Ahmed
Anjuman-I-Islam's M. H. Saboo Siddik
College of Engineering

6.1 INTRODUCTION

Passive methods for controlling vibration are not suitable for low-frequency sound and vibration as they increase the weight and volume of structures. A workable solution to this is active noise and vibration control, especially when using smart materials like piezoelectric. When employing piezoelectric actuators and sensors for active vibration control, it is essential to select the optimum size and position of the devices in order to maximize effectiveness. The beam strain energy was used as the criterion by Crawley and de Luis [1], who looked into the optimization problem in smart structures, to determine the optimal location for a piezoelectric actuator for a cantilever beam. Algorithms for actuator positioning and sizing for vibration attenuation in uniform beams were described by Devasia et al. [2]. For the purpose of determining the optimum location and size of piezoelectric actuators, they developed a number of closed-loop performance criteria. Aldraihem et al. [3] created an objective function based on beam modal cost and controllability index to determine the ideal size and location of a piezoelectric actuator/sensor. For a single pair and two pairs of actuators, the optimum size and location for the beams with various boundary conditions are discovered. Linear Quadratic Regulator (LQR) performance was used by Kumar and Narayanan [4] to discover the best location for piezoelectric actuator/sensor pairs on a beam structure. The genetic algorithm was used to solve the formulated optimization problem, which was in the form of a zero-one optimization problem. Zhang et al. [5] explored the active vibration management of beam using smart material by utilizing the LQR optimal control theory. When using a multiple-input and multiple-output (MIMO) control system to attenuate multimode vibration, linear quadratic control methods are the best choice. Zhang et al. [6] explored Linear, Quadratic, Gaussian (LQG) and H-optimal controllers for the active vibration control of a beam, and the results obtained using both controllers are contrasted.

Two approaches for active vibration control have been developed by different researchers, as can be observed from the literature for locating actuator/sensor pairs. They are either controller-dependent or controller-independent. This study adapts a controller-independent technique to determine the optimal placement of actuator/sensor pairs on beam structures using the beam strain equation. To solve the optimal location problem, a binary-coded genetic algorithm has been developed. After determining the ideal location for actuators and sensors, a state-space technique is used to build a mathematical model of the system. LQR, LQG, and H-∞ optimal controllers are designed using the full and reduced model. Using optimal controller, the simulation study of active vibration control is conducted, and the performances of each are compared.

6.2 IDENTIFICATION OF OPTIMAL LOCATION OF ACTUATOR/ SENSOR USING THE GENETIC ALGORITHM

The criterion considered for placement is the maximum strain location which is based on strain equation (curvature equation) of beam for controlling multimode vibration. This strain equation is obtained by double differentiating displacement eigen function as follows:

$$W_i(x) = C_n\{[\cos \mathbf{h}(\beta_i \times x) - \cos(\beta_i \times x)] - \alpha_i[\sin \mathbf{h}(\beta_i \times x) - \cos(\beta_i \times x)]\} \quad (6.1)$$

where $\alpha_i = \dfrac{\sin \beta_i L + \sinh \beta_i L}{\cos \beta_i L + \cosh \beta_i L}$, x = distance from fixed end, C_n = constant

$$= \frac{1}{\sqrt{\rho AL}}, i = 1 \text{ to } 6$$

$$W_i(x)'' = (C_n \times \beta_i^2)\{[\cosh(\beta_i \times x) + \cos(\beta_i \times x)] - \alpha_i[\sinh(\beta_i \times x) + \sin(\beta_i \times x)]\}$$

$$(6.2)$$

In this study, four pairs of actuators/sensors are considered for controlling multimodes of vibration. These four patch locations are determined by using the genetic algorithm because in dealing with problems involving placement of piezoelectric actuators/sensors, the genetic algorithm is considered to be the most effective optimization technique [7]. Since all the modes have the maximum strain value at the root of the cantilever beam, the strain function of all modes is algebraically added as it is considered that the present system is a linear system. Then, the combined equation can be described as follows:

$$W" = \sum_{i=1}^{n} W_i(x)" \tag{6.3}$$

here

$$W_1(x)" = (C_n \times \beta_1^2)\{[\cosh(\beta_1 \times x) + \cos(\beta_1 \times x)] - \alpha_1[\sinh(\beta_1 \times x) + \sin(\beta_1 \times x)]\} \tag{6.4}$$

$$W_2(x)" = (C_n \times \beta_2^2)\{[\cosh(\beta_2 \times x) + \cos(\beta_2 \times x)] - \alpha_2[\sinh(\beta_2 \times x) + \sin(\beta_2 \times x)]\} \tag{6.5}$$

$$W_n(x)" = (C_n \times \beta_n^2)\{[\cosh(\beta_n \times x) + \cos(\beta_n \times x)] - \alpha_n[\sinh(\beta_n \times x) + \sin(\beta_n \times x)]\} \tag{6.6}$$

The modified objective function given in equation (6.3) is to maximize, which is subjected to the following two constraints: (i) $0 \leq x \leq l$ and (ii) $loc2 - loc1 \geq 0.05$.

Utilizing strain function and a genetic algorithm, the optimal location for piezoelectric actuators and sensors is determined. Since the objective function is a single-variable function, and for a single-variable function, the genetic algorithm provides only one best optimum result, getting four optimal locations by using equation (6.3) as objective function is a difficult task. In order to determine four maximum strain locations for the current objective function, the genetic algorithm is executed in segments over the entire length of cantilever beam. The results obtained by the genetic algorithm are 0.00, 0.1670, 0.3420, and 0.4320 mm. The first location obtained using the genetic algorithm has been shifted 5 mm away from the root of beam in order to avoid practical difficulties in mounting of actuators/sensors at root, which means that the starting point of the actuator/sensor is at a distance of 5 mm, and the center point is 40 mm from the root of the beam. After positioning piezoelectric actuators/sensors in the ideal location determined by a genetic algorithm, a system model is constructed, and ANSYS© is used to carry out a modal analysis. The state-space model is then developed using MATLAB©, which will be covered in the next section.

6.3   IDENTIFICATION OF SYSTEM MODELING

Using eigenvalues and eigenvectors, a state-space technique is used to develop a dynamic model of a cantilever beam. By performing a modal analysis on a cantilever beam in ANSYS©, the eigenvalues and eigenvectors are obtained.

6.3.1   MODAL ANALYSIS

A cantilever beam was constructed in the ANSYS© software and simulated with multiple sensors and actuators. The first 10 modal frequencies and mode shapes were extracted using modal analysis. The sensor/actuator and beam employed for modal analysis have the following dimensions: beam of (length: 508 mm, width: 25.4 mm, and thickness: 0.8 mm) and piezo patch of (length: 76.2 mm, width: 25.4 mm, and thickness: 0.305 mm). At four different places on the beam, multiple pairs of actuator/ sensor are positioned together, and the optimal locations of the patches are where the strain energy is maximum as obtained by the genetic algorithm. Solid 45 element type was used for the beam and solid 5 for the piezoelectric patch. A mesh size of 70, 4, and 1 was chosen after doing a mesh convergence study. The integrated structure's modal analysis was carried out in ANSYS©, and the extracted eigenvalues for the first 10 ranks are 3.06, 18.01, 50.47, 100.62, 172.01, 287.58, 364.90, 503.77, 666.66, and 824.85. The tip node was the only node of concern because input and output were located at the beam's tip. It is possible to keep the modal matrix row that belongs to the tip node. The mass-normalized eigen-vector $Xn = [-8.625\ 9.576\ -9.395\ 9.219\ 12.51\ 11.258\ 23.603\ 12.406\ -11.597\ 8.232]$ corresponds to the beam's tip. Using these eigenvalues and eigenvectors, state-space models were created in MATLAB©.

6.3.2 State-Space Modeling

The cantilever beam mathematical model was created in MATLAB, with an impulse force as the system's tip input and tip deflection as the system's output. The input output state equations are as follows:

$$\dot{x} = Ax + Bu \tag{6.7}$$

$$y = Cx + Du \tag{6.8}$$

The system matrices A, B, C, and D are utilized to construct the system's state space. State-space models can be made using MATLAB©'s 'ss' function. By selecting the modes of frequencies with the highest dc gain values, which contribute the most to the total response, the model is reduced.

In Figure 6.1, the full model result is overlapped with the reduced model result to show the differences. For the present study, five highest contributing modes are taken to construct reduced model of the system. It is evident from Figure 6.1 that

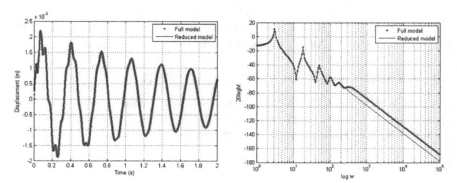

FIGURE 6.1 Transient and frequency response.

the responses of the system's reduction model closely match the responses of the complete model.

6.4 PERFORMANCE OF SYSTEM USING OPTIMAL CONTROLLERS

Determining the control signals that will cause a process to satisfy physical restrictions while also minimizing (or maximizing) some performance criterion is the goal of optimal control theory.

6.4.1 LINEAR QUADRATIC REGULATOR (LQR) OPTIMAL CONTROLLER

The following quadratic cost function was the objective for the current study's LQR controller in order to establish the optimal controller gain:

$$J = \frac{1}{2} \int_0^\infty \left(x^T Q x + u^T R x \right) dt \tag{6.9}$$

Here, the control force to be applied was u, and the suitable positive semi-definite weighting matrices Q and R were used. State feedback and output feedback are two control rules that can be utilized to develop controllers. The trade-off between the requirements for the smallness of the state and the requirements for the smallness of the control force determined the relative magnitude of Q and R in the state feedback controller rule. The form of the controlling force is as follows:

$$u = -Kx \tag{6.10}$$

The optimal controller gain, K, is as follows:

$$K = R^{-1} B^T P \tag{6.11}$$

The following algebraic Riccati equation has a unique symmetric positive semi-definite solution represented by P:

$$PA + A^T P + C^T C - PBR^{-1} B^T P = 0 \tag{6.12}$$

Using LQR function in MATLAB©, the gain (k) of the LQR controller for state feedback was calculated. >>[K,P,E]=lqr(A,B,Q,R);

Q and R were fixed as I and 1e-7 in this investigation, respectively.

State feedback equation was substituted in the system dynamic equation to get the closed-loop dynamics.

$$\dot{x} = (A - BK)x + Bu \tag{6.13}$$

The close loop of plant was constructed in MATLAB© as follows: >> sysCL=ss $(A - B * K, B, C, D)$;

The output $y(t)$ was employed in output feedback controller law instead of the state vector $x(t)$, which was part of the objective function for minimization. This could be because some state variables are not physically understood, making it more difficult to give them weight, or because the desired performance targets were better defined in terms of the measured output [8]. The following is the LQR output feedback law's algebraic Riccati equation:

$$PA + A^T P - PBR^{-1}B^T P + \left[Q - SR^{-1}S \right] \qquad (6.14)$$

$\left[Q - SR^{-1}S \right]$ was a positive semi-definite matrix.

However, MATLAB has the function lqry that simply needs the plant coefficient matrices A, B, C, and D, as well as the output and control weighting matrices, Q and R, respectively, to solve the output-weighted linear, quadratic optimum control problem as follows: >>$[K,P,E] = $lqry$(A,B,C,D,Q,R)$;

In MATLAB©, the close loop plant was created as follows: >>sysCL $= ss(A - B*K,B,C,D)$;

Figure 6.2 shows the transient and frequency responses of the LQR controller with state feedback and output feedback laws, respectively.

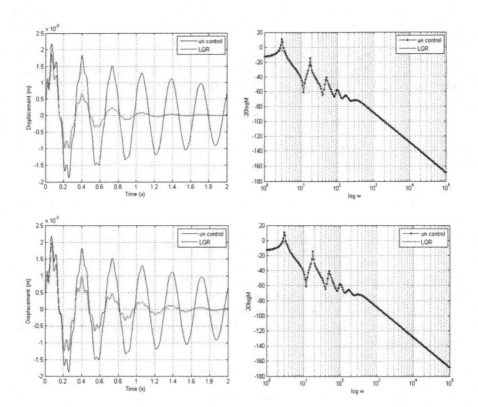

FIGURE 6.2 Transient and frequency response with state feedback and output feedback for LQR Controller.

6.4.2 LINEAR, QUADRATIC, GAUSSIAN (LQG) OPTIMAL CONTROLLER

Based on a linear plant, a quadratic objective function, and the presumption that white noise has a Gaussian probability distribution, the following can be used to represent LQG compensators in state-space form:

$$\dot{x} = Ax + Bu + Fv \tag{6.15}$$

$$y = Cx + Du + z \tag{6.16}$$

where v and z represent the process noise vector and measurement noise, respectively. The regulator gain K was calculated using equation (6.11) by solving the algebraic Riccati equation (6.12) and minimizing the LQR cost function equation (6.9) [9]. Two white noises with known power spectral densities (v and z) and known control inputs (u and y) were used to build the Kalman filter for the plant. The following state and output equations were used to create a state-space realization of the optimal compensator for controlling a noisy plant that makes use of the state-space representation of equations (6.15) and (6.16):

$$\dot{x} = (A - BK - LC + LDK)x + Ly \tag{6.17}$$

$$u = -Kx \tag{6.18}$$

Here, K stands for LQR optimum regulator gain and L for the gain matrices of the Kalman filter.

In MATLAB©, the Kalman gain L was calculated by using the 'kalman' or 'lqe' functions.

>>[est, L, E] = Kalman (sysp, v, z); >> [L, P, E] = lqe (A, B, C, v, z);

Here, 'sysp' is the state space of the plant, and v and z are the noises. In the current study, v and z are tacked as ρBB^T and $C^T C$, respectively, and ρ is a constant number. K and L for the LQG optimum regulator were determined in MATLAB using the 'reg' function: >>lqgreg = reg (sysp, K, L);

In MATLAB©, the close loop of the plant was created as: >>sysCL = feedback (lqgreg, sysp);

The regulator gain for a LQG controller with output feedback was calculated with the aid of the previously discussed LQR output controller rule. It was utilized to build the state-space model of the plant. The C output matrix became $C \times A$, and the D direct transmission matrix became $C \times B$. The transient and frequency responses of the LQG controller with state feedback and output feedback laws are shown in Figure 6.3 correspondingly.

6.4.3 H-∞ CONTROLLER

The robustness issue is immediately addressed by the H-optimal control design technique. In place of the presence of noise in the system, it produces controllers that keep system responsiveness and error signals well within set tolerances. A state-space

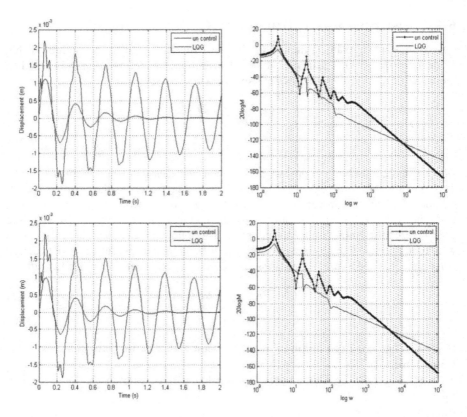

FIGURE 6.3 Transient and frequency response with state feedback and output feedback for LQG Controller.

form system's constant state feedback control is $u = -k_s x$. The state-variable feedback (SVFB) gain K_s must be identified in order for the closed-loop system $A_{cs} = A - BK_s$ to become asymptotically stable.

The state feedback gain K_s is

$$K_s = R^{-1}\left(B^T P + L\right)$$ (6.19)

Here, $P > 0$, $P^T = P$ is a solution of the following algebraic Riccati equation:

$$PA_{cs} + A_{cs}{}^T P + \frac{1}{\gamma^2} PDD^T P + Q + K_s{}^T RK_s = 0$$ (6.20)

The state-space close loop of plant with H-infinity controller was constructed in MATLAB© by using the following MATLAB© function:

```
>>[gamaopt,syscl,sysc] = hinfopt(augtf(sysp,w1,w2,w3));
```

sysp is the state-space of plant and $w1 = \dfrac{8 \times 10^{-5}\left(s^2 + 100s + 1500\right)}{0.01\left(s^2 + 2s + 1\right)}$,

$$w2 = \text{constant, and } w3 = [\ \ ].$$

The regulator gain K was calculated in MATLAB© using the 'hinfsys' or 'missys' functions along with output feedback. >>[K, sysCL, GAM]=hinfsys (P);
    where >>P=AUGW (G,w1,w2,w3); >>[K,sysCL,GAM]=mixsys (G,w1,w2,w3);
    The close loop plant with H-∞ optimal controller was constructed in MATLAB© as follows:

$$>> \text{sysCL} = \text{lft } (P, K); \text{ OR } >>[\text{GAM, acp, bcp, ccp, dcp, acl, bcl,}$$
$$\text{ccl, dcl}] = \text{hinfopt } (P); >>\text{sysCL} = \text{ss(acl,bcl.ccl,dcl)};$$

Figure 6.4 depicts the transient and frequency response of a closed-loop system with output feedback and state feedback laws, respectively.

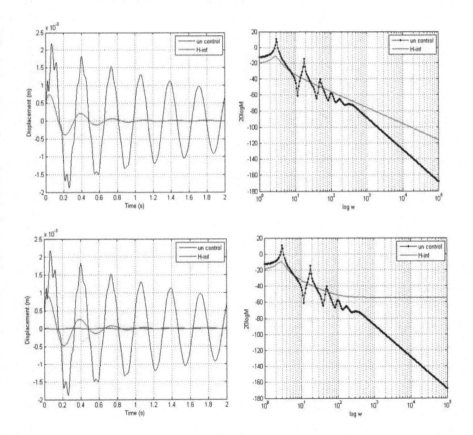

FIGURE 6.4 Transient and frequency responses with state feedback and output feedback for H-∞ Controller.

6.5 RESULTS

Figure 6.5 shows the findings of a comparison of the transient and frequency responses for state feedback and output feedback laws for the performance of the selected optimal controllers. The transient responses of all three controllers with state feedback and output feedback laws demonstrate that the settling times in all scenarios are the same. The only benefit of output feedback is that the control gain is applied directly to the sensor output. Therefore, this method reduces the internal complexity of the controller since there is no estimation of state variables required. Because the output of the system (rather than the state) possessed less information about the state, the output feedback curve is less smooth than the state feedback curve. It is clear from the transient response that the H-∞ controller performs better in a closed-loop dynamic environment than the LQR and LQG controllers.

From the frequency response, it can be inferred that the amplitude of the LQR controller matches the amplitude pattern of the uncontrolled frequency response, whereas the amplitudes of the LQG controller have been significantly reduced and the peaks of the H-infinity controller have all been eliminated, indicating much better vibration control.

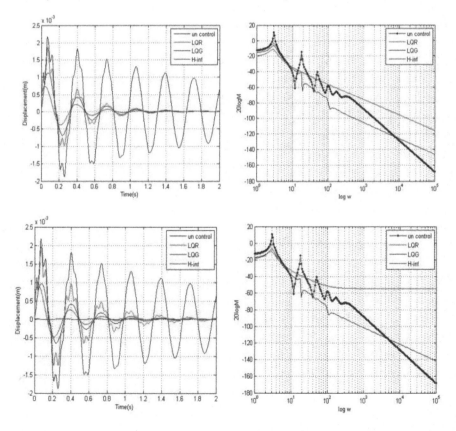

FIGURE 6.5 Transient and frequency responses with state feedback and output feedback for all the Controllers.

6.6 CONCLUSION

This study investigates active vibration control utilizing three optimal controllers (LQR, LQG, and H-∞) with state feedback and output feedback laws. In an effort to employ a genetic algorithm to locate multiple piezoelectric patches for vibration control, the strain function of a cantilever beam is used as the optimization objective function. Using the location selected by the genetic algorithm, finite element modeling of the cantilever beam integrated with multiple pairs of piezo patches is carried out. ANSYS© employs these models to carry out modal analysis. The full and reduced state-space models are constructed, and they are used to analyze the vibration response. The selected three optimum controllers with state feedback and output feedback laws are designed using the full and reduced models. The H-∞ controller outperforms the LQR and LQG controllers in a closed-loop dynamic environment, according to comparisons between the simulated results. The H-∞ controller responds more quickly than other controllers because it is designed to withstand the worst-case scenario while still achieving the necessary performance and robustness objectives. The amplitudes of the LQR controller mimic the uncontrolled frequency response, the LQG controller's amplitude has been greatly lowered, and the H-∞ controller's amplitude tends to remain constant at higher frequencies in the frequency response of the three controllers.

REFERENCES

[1] Crawley, E.F., and de Luis, J., "Use of piezoelectric actuators as elements of intelligent structure", *AIAA Journal*, Vol. 25, No. 10, 1985, pp. 1373–1385. https://doi.org/10.2514/3.9792

[2] Devasia, S. et al, "Piezoelectric actuator design for vibration suppression: Placement and sizing", *Journal of Guidance, Control and Dynamics*, Vol. 16, No. 5, 1993, pp. 859–864. https://doi.org/10.2514/3.21093

[3] Aldraihem, O.J. et al, "Optimal size and location of piezoelectric actuator/sensor: Practical consideration", *Journal of Guidance, Control and Dynamics*, Vol. 23, No. 3, 2000, pp. 509–515. https://doi.org/10.2514/2.4557

[4] Kumar, K.R. and Narayanan, S., "Active vibration control of beams with optimal placement of piezoelectric sensor/actuator pairs", *IOP Publication*, Vol. 17, No. 5, 2008, pp. 15–29. https://doi.org/10.1088/0964-1726/17/5/055008

[5] Zhang, J., He, L., Wang, E., and Gao, R., "A LQR controller design for active vibration control of flexible structures", *IEEE Pacific-Asia Workshop on Computational Intelligence and Industrial Application*, 2008. https://doi.org/10.1109/PACIIA.2008.358

[6] Zhang, J., He, L., Wang, E., and Gao, R., "Active vibration control of flexible structures using piezoelectric materials", *IEEE International Conference on Advanced Computer Control*, 2008. https://doi.org/10.1109/ICACC.2009.158

[7] Hu, J., Zhu, D., Chen, Q. and Gu, Q., "Active vibration control of a flexible beam based on experimental modal analysis", *3rd International Conference on Advanced Computer Control (ICACC 2011)*. https://doi.org/10.1109/ICACC.2011.6016429

[8] Tewari. A., *Modern Control Design with MATLAB and Simulink*, Third edition, John Wiley and Sons, West Suszex, England, 2002.

[9] Khot, S.M., Yelve, N.P., and Kumar, P. et al, Experimental investigation of performances of different optimal controllers in active vibration control of a cantilever beam. *ISSS Journal of Micro and Smart Systems*, Vol. 8, 2019, pp. 101–111. https://doi.org/10.1007/s41683-019-00040-2

7 Design and Fabrication of a Personnel Noise Enclosure for a Stone-Crushing Unit

Ansaf Mohammed Ashraf, Rohan, P.P.,
Devsuriya Devan, Nithin, R., and
Sudheesh Kumar, C.P.
Government College of Engineering

7.1 INTRODUCTION

Drilling, blasting, serial crushing, etc. are the most common activities that lead to the high-level noise and can cause adverse effects on operators and continuous listeners. Noise pollution in the crushing industry is one of the major environmental problems. The maximum industrial noise level recommended by the World Health Organization (WHO) is below 75 dB [1], and the Bureau of Indian Standards (BIS) recommends the acceptable noise level between 45 and 60 dB. When the threshold limit value (TLV) of sound exceeds the threshold of pain, it causes discomfort to people and causes chronic disorders. Sound at 85 dB (A) can lead to hearing loss if you listen to them for more than 8 hours and can damage your hearing faster. The safe listening time is cut in half for every 3 dB rise in noise levels over 85 dB (A). The hammering effect during the crushing process produces a large amount of noise and dust while working. This noise produced above a certain level is one of the critical problems associated with a stone-crushing unit. The level of noise induced in these machineries causes adverse effects on the operator and other workers.

Noise vulnerability of stone mining and crushing in Dwarka river basin of Eastern India is studied by Swadesh and Indrajit [2]. Their study focuses on control strategy for noise hazards related to machineries in crusher plants in highly exposed areas of Dwaraka river basin. A study on the intensity of noise and vibration and their control in crusher plant activities to enhance health and safety of workers was done by Ekanayake et al. [3]. They suggested methods to reduce the intensity of exposure by introducing a barrier. Noise exposure and hearing capabilities of quarry workers in Ghana by Charles et al. [4] showed the effect of noise on workers in quarry. These high levels of noise in close proximities were evaluated, and their prevention and control measures were discussed by Johnson et al. [5]. The nature of sound produced in industries and types of reduction techniques were discussed by Randall [6].

DOI: 10.1201/9781003402695-7

From the literature survey, it is clear that excessive noise from the stone-crushing machine causes high noise pollution and severe hearing impairment for workers. Thus, it is very important that measures are to be taken to reduce the noise pollution from the stone-crushing units. Taking these problems into consideration, the present paper studies the vulnerability of people to the noise produced in the stone-mining area and surroundings in the heavily stressed crushing area. This chapter also tries to analyze the noise produced during the crushing through experiment and suggests a mitigation strategy to reduce the noise levels by imparting a noise barrier.

7.2 NOISE-LEVEL MEASUREMENT

For analyzing the noise level, a crushing unit at Irikkur, Kannur, Kerala was selected. The dimensions of the complete crushing unit shown in Figure 7.1a are $24\,m \times 6\,m \times 6\,m$, which include transmission belts and secondary processing unit for stone size differentiation. The large-sized stones are fed into the unit from the top-using excavators. The noise levels are measured inside the unit at three different points at 1, 2, and 3 m away from crushing machine. Noise measurements are taken at no-load, partial-load, and full-load conditions. The measurements, at human ear level, are taken in the one-third octave band frequency using RION NL-42 type sound-level meter which allows for measurements in the range of $20–8000\,Hz$. Measurements are taken at human ear level. The noise spectrum for full-load condition is shown in Figure 7.1b which shows the maximum SPL as $91.3\,dB$. The measurement can only be taken at limited points inside the unit due to limited accessibility. The loads

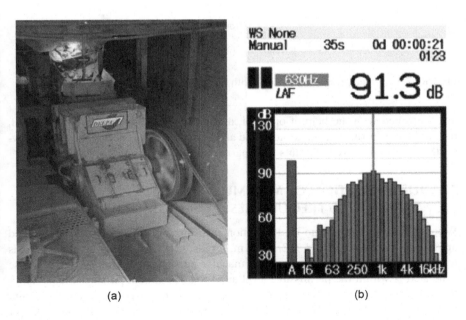

(a)   (b)

FIGURE 7.1   (a) Stone crusher and (b) sound pressure level (dB) for full load.

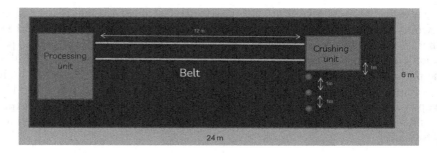

FIGURE 7.2   Schematic of the processes in the crushing unit.

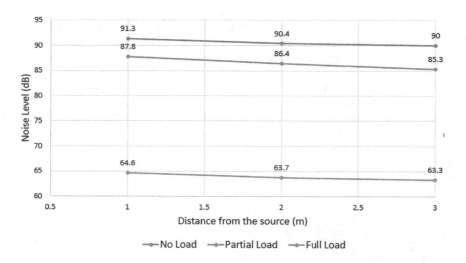

FIGURE 7.3   Sound pressure levels (dB) with different load conditions.

are controlled with the help of operator, and noise levels are measured at no-load, partial-load, and full-load conditions and at different points. The sound pressure levels (SPLs) are measured at points 1 m apart as shown in red color in Figure 7.2.

7.3   IDENTIFICATION OF MAXIMUM SPLS AND RELATED FREQUENCIES

When the crusher works at no-load condition, the sound pressure levels (SPLs) obtained are 64.6, 63.7, and 63.3 dB at 1, 2, and 3 m, respectively. The maximum noise level is found to be 64.6 dB at 1 m distance from machine as shown in Figure 7.3. So, the noise generated is within the permissible level at no-load condition. At partial-load condition, the respective maximum SPLs are 87.8, 86.4, and 85.3 dB. Here, the maximum noise level is found to be 87.8 dB at 1 m distance from machine. This possesses minor hearing problem to workers when exposed for a long time. At full-load condition, the corresponding maximum sound pressure levels are 91.3, 90.4, and 90 dB. Here, these workers are exposed to a high level of sound that has a dangerous impact on their physical and mental health. The frequency range is found to be 250–1250 Hz.

7.4 MITIGATION METHODS

There are three basic elements in any noise control system where it can be controlled:

1. The source of the sound.
2. The path through which the sound travels.
3. The receiver of the sound.

The objective of most of the noise control programs is to reduce the noise heard by the receiver. This may be accomplished by making modifications to the source, the path, or the receiver, or to any combination of these elements. In the case of crusher, it is impossible or difficult to implement a noise control system at source due to heavy-loading system and space restrictions. So, the present chapter proposes to reduce the noise by building an acoustic enclosure in the sound transmission path to the operator to protect him from harmful noise levels and dust.

7.5 EXPERIMENT

7.5.1 SELECTION OF MATERIALS

An acoustic enclosure consists mainly of two parts: a supporting basic structure and a non-transmitting layer with good sound transmission class (STC) value. STC is an integer rating of how well a building partition attenuates airborne sound. The higher the value of STC, the lesser the intensity of the sound that can be heard. For constructing the basic structure of the enclosure cabin, the material selected is plywood of 12 mm thickness which has enough structural strength and is economical. Plywood has an STC value of 23 and is relatively cheap as shown in Table 7.1. The material used for noise reduction is chosen based on the criteria that it transmits less sound and can be used in an environment which is always filled with dust.

TABLE 7.1
Sound Pressure Levels Inside the Enclosure

Material	STC Rating	Density (kg/m³)	Cost per Square Feet (Rs.)	Characteristics
Plywood (12 mm thick)	23	550	45	High strength, dimensional stability, sound insulation
Thermocol	Very low	20	12	Low strength, easily breakable
Wood (teak wood) (50 mm thick)	30	1500	2250	Complex molding process required
MLV (5 mm thick)	32	2100	190	Absorb sound waves High tear resistance

After comparing different acoustic materials like fiber glass wool, melamine foam, acoustic tiles, polyurethane foam, etc., the material selected is mass-loaded vinyl (MLV) (Figure 7.4a). MLV is an effective, flexible, and affordable product for soundproofing when used correctly. It has an STC value of 32 and costs Rs. 190 per square feet as given in Table 7.1. On studio projects, movie theaters, and residential applications, it can improve the STC rating of a wall and help attenuate excess noise. The transmission loss at different frequencies of MLV is depicted in Figure 7.4b which shows that the transmission loss is high in the frequency of 500–2000 Hz, and the frequency of noise to be reduced is within this range. Hence, the material is very appropriate for the construction of the enclosure cabin.

7.5.2 Fabrication

An acoustic enclosure is constructed using the material selected as discussed in mitigation strategy, with two materials, plywood of 12 mm for basic structure, and MLV of 5 mm coated as non-transmitting material. A miniature of a practical cabin is constructed with a dimension of 1 m × 0.6 m × 0.8 m which is one-third of an actual

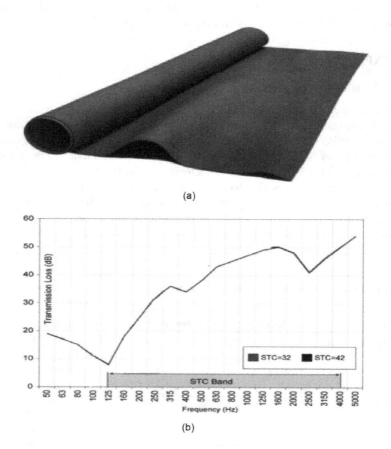

(a)

(b)

FIGURE 7.4   (a) Mass-loaded vinyl (MLV) and (b) transmission loss of MLV.

cabin which requires a size of 3 m × 1.8 m × 2.4 m. This structure gives enough space to keep the sound sensor of sound-level meter inside the enclosure, and a small opening is provided to take the extension wire inside for measurement of sound pressure level. The construction of basic structure requires two sheets of plywood, a 4 × 8 ft and a 4 × 6 ft sheet which cost a total of Rs. 2500 as shown in Figure 7.5.

7.5.3 SOUND SOURCE

The source used in the lab is BOAT stone 40 W Bluetooth speaker which can produce SPL of desired amplitude and frequency. The amplitude of maximum sound level is experimentally found around a constant distance of 0.7 and 1.4 m at 630 Hz, which is almost the same as the frequency with which the maximum noise is found in the crusher. The source is kept in human ear level, and the measurements were taken in the same level. The result is shown in Figure 7.6, and the variations around the source are due to the variation in reflection from the walls around.

7.5.4 MEASUREMENT

To find the amount of reduction in noise inside the enclosure, measurement is made using RION sound-level meter inside the enclosure by hanging the sensor at 0.5 m

FIGURE 7.5 Enclosure for lab experiment.

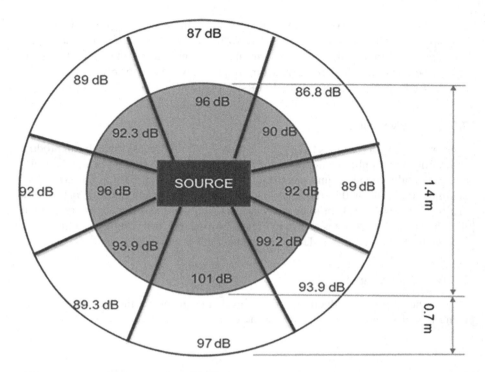

FIGURE 7.6 Sound pressure levels (dB) around the source.

above the ground, at its center. The measurement of sound pressure level was done at a position 1 m away from the source without any enclosure or barrier at 630 Hz, and the amplitude was found 93.7 dB by adjusting the volume which is in the same range as the maximum nose obtained in the crushing unit. The sound levels at different frequency were also found at the same point without changing the amplitude. The results are noted and used to compare the noise levels at the receiver end in the presence of enclosure with and without MLV layer coating. Then, the measurements are made at different frequencies inside the enclosure at 1 m distance without changing the amplitude of sound. The measurements are taken for three cases, i.e., with (i) no enclosure, (ii) plywood enclosure, and (iii) MLV enclosure. The results are given in Table 7.2 and Figure 7.7.

It is found that there is a significant reduction in noise when a 12 mm plywood enclosure is used. Further reduction is achieved when coated with an acoustic MLV of 5 mm thickness. This reduction is comparatively high in the frequency range of 500–1250 Hz. This is the required range of frequency of SPLs to be reduced in the crushing unit. The results show that even with just plywood enclosure, the noise can be reduced significantly at some frequencies. A very high reduction in SPL of about 23 dB is found with enclosure made of plywood and coated with MLV material in the frequencies of 200 and 1250 Hz.

TABLE 7.2
Sound Pressure Levels Inside the Enclosure

	Sound Pressure Levels (dB)		
Frequency (Hz)	No Enclosure	Plywood Enclosure	MLV Enclosure
1250	98.5	86.8	75.0
1000	91.7	84.0	77.9
800	96.7	90.0	82.0
630	93.7	83.9	75.2
500	89.3	86.0	73.0
400	83.7	74.6	69.7
315	78.8	72.3	68.2
250	91.3	73.4	65.7
200	89.2	71.0	66.0

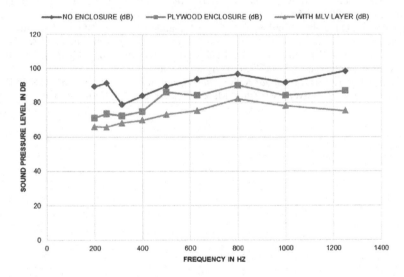

FIGURE 7.7　Sound pressure levels with and without enclosure.

7.6　CONCLUSIONS

Measurement of SPLs in crushing units carried out in this chapter shows that the workers are exposed to noise levels as high as 93 dB. A noise enclosure for crushing unit operators is designed and fabricated in the present work. The results show that the proposed enclosure can significantly minimize the SPLs experienced by the workers. It is found that a simple cabin made of plywood with 12 mm thickness can reduce the sound significantly at some frequencies. Thus, all crushing units can at least have a cabin made of plywood so that the noise levels, the workers are exposed to, can be reduced to some extent. A coating by MLV over the plywood is found to reduce the SPL by 23 dB in the lowest and highest frequencies. Also, it is not too costly to design and fabricate such a personnel enclosure in crushing units.

REFERENCES

[1] Berglund, B., Lindvall, T., and Schwela, D.H. New who guidelines for community noise. *Noise & Vibration Worldwide*. 2000;31(4):24–29. doi:10.1260/0957456001497535

[2] Pal, S., and Mandal, I. Noise vulnerability of stone mining and crushing in Dwarka riverbasin of Eastern India. *Environment Development and Sustainability*. 2021;23:13667–13688. doi:10.1007/s10668-021-01233-2

[3] Thiruchittampalam, S., Kinoj, A., Ekanayake, E., Vithurshan, S., Hemalal, P., Samaradivakara, G., Rohitha, L. and Chaminda, S. Noise and vibration control in crusher plant activities to enhance health and safety of workers. *Conference: International Symposium on Earth Resources Management and Environment. Grand Monarch Thalawathugoda, Sri Lanka.*

[4] Gyamfi, C.K.R. Noise exposure and hearing capabilities of quarry workers in Ghana: A cross-sectional study. *Journal of Environmental & Public Health*. 2016;2016(11):1–7. doi:10.1155/2016/7054276

[5] Johnson, D.L., Papadopulos, P., Watfa, N., and Takala, J. (1998). Exposure criteria, occupational exposure levels. In: B. Goelzer, C.H. Hansen, & G.A. Sehrndt, *Occupational Exposure to Noise: Evaluation, Prevention, and Control*, pp. 79–102. Geneva, Switzerland: World Health Organization.

[6] Randall, F. (2001). *Industrial Noise Control and Acoustics*. Louisiana, U.S.A: Barron Louisiana Tech University Ruston, Marcel Dekker, Inc. ISBN: 0-8247-0701-X.

8 Investigation of Effect of Porous Material on Performance of Helmholtz Resonator

Nilaj N. Deshmukh, Afzal Ansari,
and Axin A. Samuel
Agnel Charities' Fr. C. Rodrigues Institute of Technology

8.1 INTRODUCTION

Instabilities caused by thermo-acoustic oscillations have been observed in various kinds of combustion systems. Phenomena related to the thermo-acoustics have also been seen in other systems that have heat transfer processes. For example, glass blowers have reported observations of spontaneous excitation of acoustic oscillations during the heating of the closed end of a blown-glass tube [1]. A similar phenomenon has also been observed in cryogenic storage vessels, where oscillations are induced by the insertion of a hollow tube open at the bottom end in liquid helium. In both cases, there is a thermal source that causes acoustic oscillation. Due to this, the resulting instability is referred to as thermo-acoustic instability (TAI).

Thermo-acoustic instabilities are self-excited by a feedback loop between unsteady heat release and one of the natural acoustic modes of the combustor [1]. Combustion instabilities are undesirable in most cases, as they are characterized by unsteady heat release and pressure oscillation. Pressure fluctuations are dangerous as they can lead to excessive vibration resulting in mechanical failures, high levels of acoustic noise, high burn and heat transfer rates, and possibly component melting [2]. In some cases such as ramjets and pulsed combustors, engines inherently depend on the presence of sustained oscillation [3] (Figure 8.1).

The Rijke tube is a prototype to study the phenomenon of TAI on a laboratory scale [4–6]. It consists of heated wire gauze or a burner at a distance $L/4$ from one end within it, which acts like a source of sound. The intensity of sound is proportional to input power [6]. Control of TAI can be achieved by either passive or active methods. Active methods consist of an external perturbation which is introduced to the system via an actuator, which manipulates the acoustic field or the heat injection rate. Active control methods are more 'adaptive' and have great potential to stabilize combustors even at off-design conditions [7]. Active control methods can be subdivided into two methods: open-loop active control methods and closed-loop active

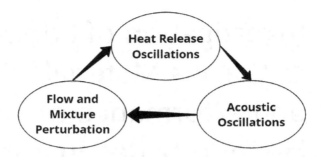

FIGURE 8.1   Illustration of the feedback processes responsible for combustion instability.

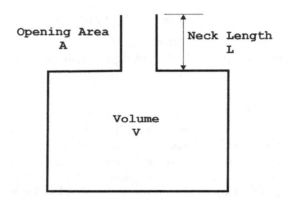

FIGURE 8.2   Schematic diagram of Helmholtz resonator.

control methods. Various open-loop control systems are use of heaters, swirlers, jet in crossflow, etc. [7,8]. Fuel modulation and flow modulation are major closed-loop control approaches to abate TAI [5,9,10].

Passive methods are those which don't involve use of power. Such methods are developed after studying entire combustion phenomenon, combustion chamber geometry, etc. These methods are simple, less expensive, and a robust all-time solution to the problem. Important passive techniques to abate TAI include use of acoustic dampers like Helmholtz resonator (HR), quarter wave tubes, perforated liners, PIMs, and baffles [11–18]. Other passive techniques involve use of miniature vortex generators, drum-like silencer, and change in combustor geometry and fuel injector system [19]. They are applicable around a certain frequency, which has led to the research of self-tuning and closed-loop Helmholtz resonator [20–23]. Use of acoustic damping materials such as granular aerogel and fibrous materials inside Helmholtz resonators has also been recently studied [24–26].

A Helmholtz resonator is a widely used acoustic damper. It consists of short neck attached to a cavity as shown in Figure 8.2. This study focuses on improving the HR with help of porous materials. An inexpensive and readily available porous material is selected, and the surface opposite to the neck of the HR is lined with it. The performances of the HR with and without porous material are then compared.

8.2 DESIGN AND FABRICATION OF HELMHOLTZ RESONATOR

8.2.1 DESIGN OF ADAPTIVE PASSIVE HELMHOLTZ RESONATOR

The Helmholtz resonator is designed to dampen the second thermo-acoustic mode characterized within the Rijke tube of length 800 mm and diameter 80 mm which is attached to a plenum chamber. The thermo-acoustic instability lies within the range of 400–500 Hz.

The designing process of the Helmholtz resonator consists of the following calculations:

$$f = \frac{c}{2\pi} \sqrt{\frac{S_d}{L_n \times V}} \qquad (8.1)$$

where
f = frequency (measured experimentally)
C = speed of sound
S_d = neck area
L_n = length of neck
V = volume of resonator cavity

The area of neck is calculated as follows:

$$S_d = \frac{\pi (d_n)^2}{4} \qquad (8.2)$$

where d_n = diameter of the neck.

The volume of resonator cavity is given by

$$V = \frac{\pi \times (D_c)^2 \times L_c}{4} \qquad (8.3)$$

where
D_c = diameter of resonator cavity
L_c = length of Resonator cavity

The resonator consists of two parts, viz. the cavity and the cap. The cavity consists of a cup-like structure with the lip of the cup having internal threading and a hole at the bottom with the neck of the resonator attached to it, and the neck has external threading on it to make it easier to attach it to the Rijke tube. The cap consists of a smooth plug-like structure with a flange on top, and the plug has a bit of external threading which meshes with the internal threading of the cup, as shown in Figure 8.3. The pitch of the thread is 2 mm so each rotation of the flange varies the height of the cavity by the same and thus, changes the volume of the cavity inside the resonator.

The diameter of the cavity is 40 mm, and the height of the cavity can be varied from 5 to 15 mm. The neck diameter of the resonator was taken as 6 mm and the length of

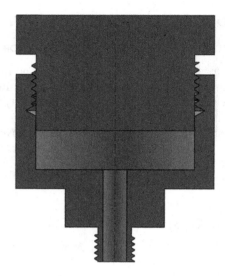

FIGURE 8.3   CAD model of cross-section of resonator assembly.

the neck as 24 mm. The optimum cavity height for suppression of thermo-acoustic instability of the second mode for this particular Rijke tube was found to be 11 mm.

8.2.2   FABRICATION OF HELMHOLTZ RESONATOR

The parts of the resonator were made using 50 mm diameter mild steel stock. The cup and the cap were both machined on a lathe. To manufacture the cup, first a hole of 6 mm for the neck was drilled through the material. Next, the hole was further enlarged enough near the cavity to fit the boring tool, and boring process was carried out until the inner diameter of the cup reached 40 mm. Threads were cut on the inner surface of the cup till a depth of 15 mm. The neck section was then machined, and M10 threads were cut on it. The cap was similarly machined to fit the cup, and diamond knurling was done on the flange to make it easier to vary the height of the cavity.

8.2.3   SELECTION OF POROUS MEDIA

Polyurethane foam with 1.45 PCF (pounds per cubic foot) was used due to its cheap cost and ease of shaping it to fit inside the Helmholtz resonator cavity. The foam was a 0.5 mm thick circular sheet glued to the cap as shown in Figure 8.4b, while leaving some space between the neck and foam to prevent blocking the cavity.

8.3   TEST SETUP AND EXPERIMENTATION

The experimental setup consists of a Rijke tube having length 800 mm, internal diameter 80 mm, and thickness 3 mm and is made up of mild steel. A coaxial brass

(a)   (b)

FIGURE 8.4 (a) Resonator cavity without porous insert and (b) placement of porous material in resonator.

FIGURE 8.5 Experimental setup.

burner of 30 mm diameter, connected to an inlet pipe of diameter 7 mm can move to and fro inside the tube with the help of a traverse. It has a ceramic head with 60 holes. It is connected to a pre-mixer in which air and liquid petroleum gas (LPG) are mixed in required proportion with the help of vortex flow arrangement. The flow rates of LPG and air to the pre-mixer can be changed with the help of rotameters. The range of air rotameter is 0–10 lpm with the least count as 0.2 lpm, whereas LPG rotameter has a range of 0–0.5 lpm with the least count as 0.025 lpm. The fuel used in the present experiments is commercially available LPG.

The Rijke tube was threaded and connected to the plenum chamber to prevent any leakage of air. The other end of the plenum chamber was connected to a blower which acts as a suction pump and was used to set the mass flow rate of air flowing through the system. Figure 8.5 shows the entire experimental setup, while Figure 8.6 is the schematic of the entire setup.

MI-1432 microphone with MI-3111 pre amplifier manufactured by Ono Sokki was used to capture the instability. The microphone was connected to NI 9234 using a BNC-to-BNC cable. Honeywell AWM700 Series microbridge mass airflow sensors which provide in-line flow measurement were used to measure the flow of air through the system.

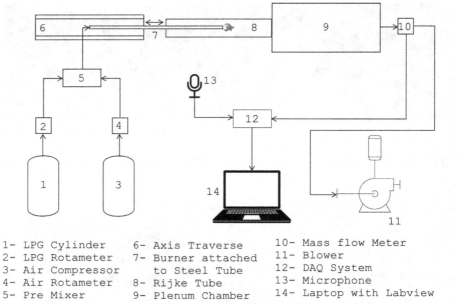

1- LPG Cylinder	6- Axis Traverse	10- Mass flow Meter
2- LPG Rotameter	7- Burner attached	11- Blower
3- Air Compressor	to Steel Tube	12- DAQ System
4- Air Rotameter	8- Rijke Tube	13- Microphone
5- Pre Mixer	9- Plenum Chamber	14- Laptop with Labview

FIGURE 8.6 Schematic of experimental setup for studying TAI in Rijke tube.

8.3.1 EXPERIMENTAL PROCEDURE

The experimentation process consists of four parts:

- Characterization of thermo-acoustic instability
- Finding the antinode
- Designing the Helmholtz resonator
- Final experiment

Characterization of thermo-acoustic instability consists of capturing the points inside the Rijke tube where the instability is induced. The frequency of thermo-acoustic mode is a function of power input, air-to-fuel ratio (AFR), and burner position along the longitudinal axis of the Rijke tube. Keeping these parameters fixed, TAI is made to appear inside the Rijke tube. The characterization process is done by fixing a particular flow rate of air though the tube and then inserting the burner in increments of 20 mm while capturing the acoustic signal. Figure 8.7 shows the block diagram used for that particular VI.

The antinode is found by calculating the fundamental frequency of the system by using the following formula:

$$f = \left(\frac{c}{2L}\right) \tag{8.4}$$

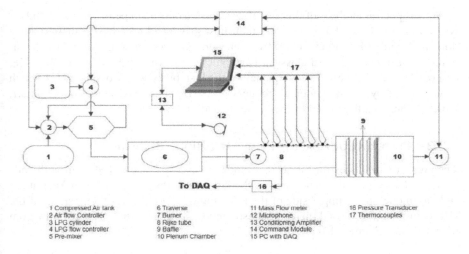

1 Compressed Air tank	6 Traverse	11 Mass Flow meter	16 Pressure Transducer
2 Air flow Controller	7 Burner	12 Microphone	17 Thermocouples
3 LPG cylinder	8 Rijke tube	13 Conditioning Amplifier	
4 LPG flow controller	9 Baffle	14 Command Module	
5 Pre-mixer	10 Plenum Chamber	15 PC with DAQ	

FIGURE 8.7 Block diagram used for the VI.

where

c = speed of sound

L = length of tube

The mode can then be found by dividing the frequency of the instability by the fundamental frequency. The antinode can then be found for the corresponding mode.

The designing process of the Helmholtz resonator has been described in Section 8.2.1.

The final experimentation step of the process consists of using the newly designed resonator to try to suppress the instabilities that were found during characterization. The raw data were first captured using the microphone and then converted to frequency domain using fast Fourier transform (FFT). The data obtained were further converted to sound pressure level (SPL) (dB) using the following equation:

$$10\log_{10}\left[\frac{\Delta f \times \left(\dfrac{10^{\left(\frac{v}{10}\right)}}{\text{calibration factor}^2}\right)}{(2e-5)^2}\right] \qquad (8.5)$$

8.3.2 SECOND THERMO-ACOUSTIC MODE

Keeping the air flow rate as 9 lpm and LPG flow rate as 0.2 lpm and burner position at 42 cm from the left end, a thermo-acoustic mode was captured. The AFR was 45. The flame was blue conical premix flame.

When spectral analysis of this signal was done, a dominant peak at around 469 Hz was observed. This FFT signal was then post processed to get SPL in frequency domain. Figure 8.8 denotes SPL in frequency domain. As per equation (8.4), the fundamental, second and third classical harmonics are of this Rijke tube are 219, 438, and 657 Hz, respectively. The frequency of the instability was around 469 Hz which when divided by the fundamental frequency, 219 Hz, in our case comes out to be 2.1, which means that the instability is of the second mode.

8.3.3 PLACEMENT OF ADAPTIVE PASSIVE HELMHOLTZ RESONATOR

The pressure antinodes for the second mode are at a distance of L/4 from both ends of the Rijke tube. With burner near to the left side of the tube, the HR is kept at the right antinode on the upper side of the tube. Figure 8.9 indicates the position of the HR to dampen the second thermo-acoustic mode. It lies at 848/4–24 = 188 mm from the right end (Figure 8.10).

8.4 RESULTS

Five sets of readings are taken without inserting the porous material. These correspond to cavity height changed from 8 to 12 mm in increments of 1 mm. SPL variation with frequency of these readings is given in Figure 8.11a. To check the effectiveness of the HR, curves corresponding to SPL without using the HR and the background noise are superimposed on these readings.

The same procedure is repeated for HR with porous insert. Five sets of readings are taken with the porous material insert. These correspond to cavity height changed from 8 to 12 mm in increments of 1 mm. SPL variation with frequency of

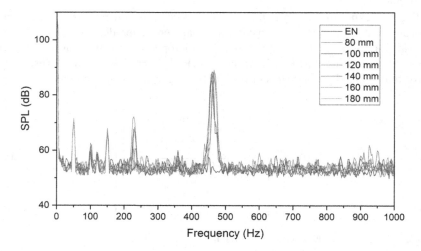

FIGURE 8.8 SPL as a function of frequency with dominant second thermo-acoustic mode across various lengths in tube.

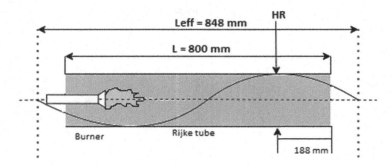

FIGURE 8.9 Placement of Helmholtz resonator.

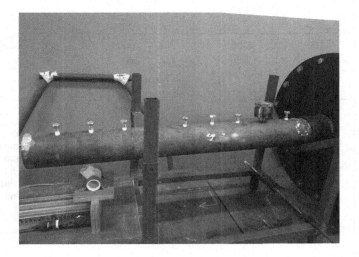

FIGURE 8.10 Helmholtz resonator screwed to the Rijke tube.

these readings are given in Figure 8.11b. It is seen that with porous material, the suppression of instability takes place at lower cavity. To get a better understanding of the results, comparison is made between readings of HR and HR with porous material as shown in Figure 8.12.

In Figure 8.12a and b, it can be seen that there is thermo-acoustic instability, and both the resonators at 8 and 9 mm cavity height are not effective. Starting from Figure 8.12c for 10 mm cavity height, it can be seen that the resonator with porous material is much more effective at suppressing the instability than the regular HR which only shows a slight suppression. With 11 mm cavity height, both HRs suppress the thermo-acoustic instability as seen in Figure 8.12d when the cavity height is further increased to 12 mm. In Figure 8.12e, it can be seen that the regular HR starts letting the instabilities back, whereas the resonator with porous material still suppresses the instability.

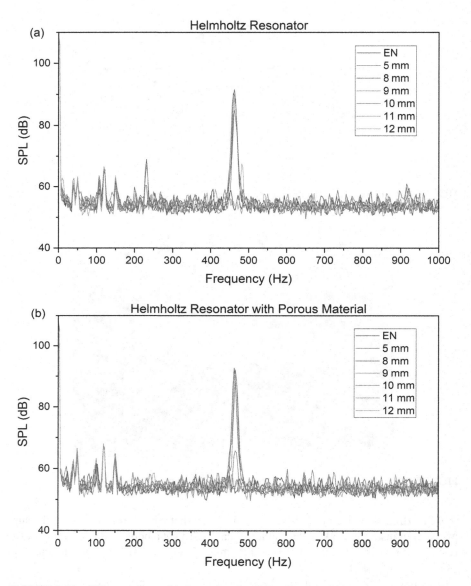

FIGURE 8.11 SPL variation with frequency for different cavity height of HR (b) with and (a) without porous material.

In Figure 8.13, SPL is plotted as a function of cavity height of HR. When cavity height is kept at the calculated height for the setup, i.e., 11 mm, instability is suppressed. It can be seen from the diagram that with the use of porous material, effectiveness of the HR is increased (decrease in SPL) (Figure 8.13).

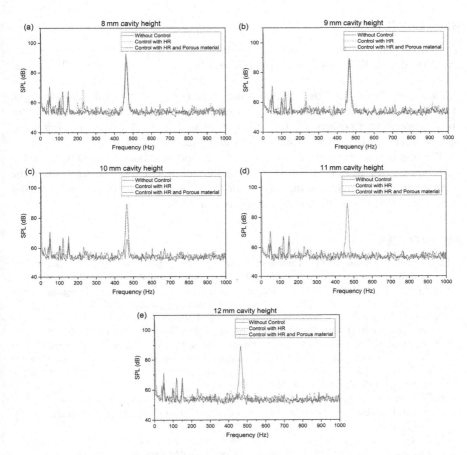

FIGURE 8.12 Comparison between HR and HR with porous insert for 12 mm cavity height.

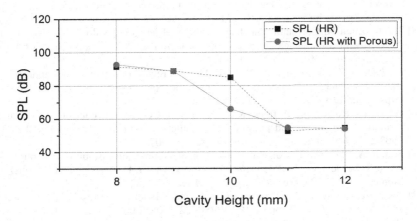

FIGURE 8.13 Effectiveness of HR at different cavity heights.

8.5 CONCLUSION

The phenomenon of thermo-acoustic instability was successfully studied and miti-gated inside the Rijke tube using the HR and HR with porous insert. The second thermo-acoustic mode corresponding to a frequency of 469 Hz was captured. Screw threads were used to vary the cavity volume of the HR by changing the cavity height. It was seen that inserting porous material reduces the volume of cavity required for suppression of thermo-acoustic instability. A reduction of 9% of cavity volume was observed when using porous inserts with Helmholtz resonator.

The research focuses on the feasibility of using polyurethane foam to improve the performance of HR. Further research is required for practical implementation of the method, and parameters like thickness of foam, density, and stiffness can be further optimized to improve damping characteristics of the resonator.

REFERENCES

[1] Zinn BT, Lieuwent TC. Combustion instabilities: Basic concepts. In *Combustion Instabilities in Gas Turbine Engines*, Eds. T. Lieuwen and V. Yang. 2012;210:3–26
[2] Poinsot T. Prediction and control of combustion instabilities in real engines. *Proc Combust Inst* 2017;36:1–28. https://doi.org/10.1016/j.proci.2016.05.007.
[3] Sun X, Wang X. Thermoacoustic instability. *Fundam Aeroacoustics Appl Aeropropulsion Syst* 2021;415–99. https://doi.org/10.1016/b978-0-12-408069-0.00007-5.
[4] Deshmukh NN, Kudachi B, Joy S, Phansalkar S, Pillai V, Thomas T. Suppression of Thermo-Acoustic Instabilities using Helhmoltz Resonator. In *International Conference on Nascent Technologies in Engineering*, Navi Mumbai, India. 2019:1–6 https://doi.org/10.1109/ICNTE44896.2019.8945994.
[5] Deshmukh NN, Sharma SD. Suppression of thermo-acoustic instability using air injection in horizontal Rijke tube. *J Energy Inst* 2017;90:485–95. https://doi.org/10.1016/j.joei.2016.03.001.
[6] Atis CAA, Sarker M, Ehsan M. Study of thermoacoustic phenomenon in a rijke tube. *Procedia Eng* 2014;90:569–74. https://doi.org/10.1016/j.proeng.2014.11.774.
[7] Mahesh S, Gopakumar R, Rahul B V., Dutta AK, Mondal S, Chaudhuri S. Instability control by actuating the swirler in a lean premixed combustor. *J Propuls Power* 2018;34:708–19. https://doi.org/10.2514/1.B36366.
[8] Tao C, Zhou H. Correlation analysis of oxy-fuel jet in cross-flow on thermoacoustic instability in a model gas turbine combustor. *Aerosp Sci Technol* 2020;106:106184. https://doi.org/10.1016/j.ast.2020.106184.
[9] Zhao D, Lu Z, Zhao H, Li XY, Wang B, Liu P. A review of active control approaches in stabilizing combustion systems in aerospace industry. *Prog Aerosp Sci* 2018;97:35–60. https://doi.org/10.1016/j.paerosci.2018.01.002.
[10] Guyot D, Rößler M, Bothien MR, Paschereit CO. Active control of combustion instability using pilot and premix fuel modulation. *14th Int Congr Sound Vib 2007* 2007;2:1283–92. https://doi.org/10.1002/pamm.2007.
[11] Frendi A, Nesman T, Canabal F. Control of combustion-instabilities through various passive devices. In *Proceeding of 11th AIAA/CEAS Aeroacoustics Conference*, Monterey, CA, 2005 https://doi.org/10.2514/6.2005-2832.
[12] Tang PK, Sirignano WA. Theory of a generalized Helmholtz resonator. *J Sound Vib* 1973;26:247–62. https://doi.org/10.1016/S0022-460X(73)80234-2.
[13] Howard CQ, Craig RA. Noise reduction using a quarter wave tube with different orifice geometries. *Appl Acoust* 2014;76:180–6. https://doi.org/10.1016/j.apacoust.2013.08.006.

[14] Park IS, Sohn CH. Nonlinear acoustic damping induced by a half-wave resonator in an acoustic chamber. *Aerosp Sci Technol* 2010;14:442–50. https://doi.org/10.1016/j. ast.2010.04.011.

[15] Eroglu S, Toprak S, Urgan O, Onur OE, Denizbasi A, Akoglu H, Ozpolat C, Akoglu E. Passive control of combustion noise and thermo-acoustic instability with porous inert media. *Saudi Med J* 2012;33:3–8.

[16] Justin Williams L, Meadows J, Agrawal AK. Passive control of thermoacoustic instabilities in swirl-stabilized combustion at elevated pressures. *Int J Spray Combust Dyn* 2016;8:173–82. https://doi.org/10.1177/1756827716642193.

[17] Ma XQ, Su ZT. Development of acoustic liner in aero engine: a review. *Sci China Technol Sci* 2020;63:2491–504. https://doi.org/10.1007/s11431-019-1501-3.

[18] Zhao D, Morgans AS, Dowling AP. Tuned passive control of acoustic damping of perforated liners. *AIAA J* 2011;49:725–34. https://doi.org/10.2514/1.J050613.

[19] Sohn CH, Park JH. A comparative study on acoustic damping induced by half-wave, quarter-wave, and Helmholtz resonators. *Aerosp Sci Technol* 2011;15:606–14. https:// doi.org/10.1016/j.ast.2010.12.004.

[20] Zhao D, Li XY. A review of acoustic dampers applied to combustion chambers in aerospace industry. *Prog Aerosp Sci* 2015;74:114–30. https://doi.org/10.1016/j. paerosci.2014.12.003.

[21] Kurdi MH, Duncan GS, Nudehi SS. Optimal design of a helmholtz resonator with a flexible end plate. *J Vib Acoust Trans ASME* 2014;136:031004. https://doi. org/10.1115/1.4026849.

[22] Birdsong CB, Radcliffe CJ. A smart Helmholtz resonator. In *Proceedings of ASME Forum Act Noise Control*, Dallas, TX, 1997:1–5.

[23] Dannemann M, Kucher M, Kunze E, Modler N, Knobloch K, Enghardt L, et al. Experimental study of advanced helmholtz resonator liners with increased acoustic performance by utilising material damping effects. Appl Sci 2018;8:1–18. https://doi. org/10.3390/app8101923.

[24] Wang W, Zhou Y, Li Y, Hao T. Aerogels-filled Helmholtz resonators for enhanced low-frequency sound absorption. J Supercrit Fluids 2019;150:103–11. https://doi. org/10.1016/j.supflu.2019.04.011.

[25] Selamet A, Xu MB, Lee I-J, Huff NT. Helmholtz resonator lined with absorbing material. J Acoust Soc Am 2005;117:725–33. https://doi.org/10.1121/1.1841571.

[26] Hong Z, Dai X, Zhou N, Sun X, Jing X. Suppression of Helmholtz resonance using inside acoustic liner. J Sound Vib 2014;333:3585–97. https://doi.org/10.1016/j.jsv.2014.02.028.

9 Application of Finite Element Method for Analyzing the Influence of Geometrical Parameters of Spur Gear Pair on Dynamic Behavior

Achyut. S. Raut
Rajendra Mane College of Engineering and Technology
Fr. C. Rodrigues Institute of Technology

S. M. Khot and Vishal G. Salunkhe
Fr. C. Rodrigues Institute of Technology

9.1 INTRODUCTION

Gears are most widely used mechanical components due to their unique technical advantages. Under certain dynamic circumstances, geared systems are known to generate a significant mesh and bearing forces. Due to their durability and noise limits, their dynamic forces are predicted throughout the design stage. For improved transmission design, the capacity to precisely compute the dynamic loads in geared systems becomes crucial. For researching the dynamics of gear systems, numerous studies have been carried out. The researchers have proposed different models to find dynamic forces on gear teeth.

In 1992, Kahraman et al. [1] developed a finite element (FE) model to examine the dynamic behavior of geared systems. Their findings demonstrated that the dynamics of geared systems are significantly impacted by bearing compliances. Nonlinear and time-varying effects are modeled in a group of gear dynamic studies. Such studies enable researchers to investigate the impacts of various gear faults, shape modifications, mounting errors, and gear dynamics. In 1994, Lin et al. [2] investigated the influence of various profile revisions on dynamic loads. In 1996, Velex and Maatar [3] illustrated the impact of mounting inaccuracies and shape abnormalities on gear dynamics. In the same year, Liou et al. [4]

DOI: 10.1201/9781003402695-9

proposed a computer simulation to demonstrate the effect of gear contact ratio on the dynamic load of a spur gear. The reviews made by Ozguven and Houser [5] and Wang et al. [6] gave widespread empirical models for deterministic gear dynamics. Those studies have significantly contributed to understanding of the dynamics and improvement of dynamic performance of gear sets. However, the significance of existing studies from deterministic perspective is very limited while dealing with vibrations caused by random excitation. In 2014, Wen et al. [7] studied the spur gear-pair dynamics in response to harmonic and white noise excitation. The authors used the path integration method to analyze reactions by considering the backlash mesh stiffness.

Researchers have recently created more sophisticated dynamic models to examine the influence of nonlinear gear characteristics on the dynamics of gear-pair systems. In 2004, Kabur et al. [8] suggested an empirical model for a helical gear that incorporates bearing and housing flexibility. They investigated the free and forced vibration of a system to illustrate the impacts of various system factors like shaft angles, bearing stiffness, and gear hand. In the same year, Maliha et al. [9] created a loaded static transmission error approach that incorporates the model excitation effects of gear faults and profile alterations. Ozguven and Houser [10] investigated the interaction of numerous distinct components considered in the analysis with backlash and shaft flexibility.

In 2007, Tamminana et al. [11] anticipated the dynamic behavior of deformable bodies based on FEs for spur gear pairs. In the same year, Li [12] illustrated an FE method for spur gear analysis with machining defects, tooth alterations, and load-sharing ratio. In 2010, Kim et al. [13] proposed a time-varying dynamic model. The investigated model is highly accurate than previous models in terms of predicting dynamic response. In 2012, Moradi and Salarieh [14] investigated spur gear-pair backlash nonlinearity and nonlinear oscillations.

In 2016, Ma et al. [15] modified addendums and tooth profiles to develop mesh stiffness for profile-shifted gears. The loaded static transmission error at modification levels is evaluated for determining the tooth profile modification curve. In 2017, Wang [16] illustrated variables such as backlash, mesh stiffness, and STE when developing a 3-DOF model of a locomotive transmission system. In 2021, Liu et al. [17] evaluated deviation in pitch on response features of a spur gear with contact ratio. In 2022, Gao et al. [18] proposed an optimization technique for an involute spur gear pair for transmission system. The analysis of literature shows that a large amount of design parameters of bearings, shafts, housing, and gears can influence the overall dynamic behavior. Numerous researchers have made significant efforts for the analysis of gear vibration and noise, and proposed methods to minimize it. Despite widespread agreement on the nature of the phenomenon, the current knowledge of gear vibration is limited. As a result, there is space to investigate the impact of distinct variables such as addendum, backlash, and modified tooth profiles by changes in the geometrical properties of gear teeth profiles.

The development of an FE approach capable of simulating the impact of considerable parametric fluctuations under conditions of constant load and speed is the objective of the current work. For that purpose, three different sets of spur gear pair are designed with variations in geometrical parameters. The dynamic response of gear pairs will be predicted using an FE method. Analysis should be carried out to identify the most significant factor which affects the dynamic behavior of spur gear pair. Finally, the results obtained by the FE method should be validated by comparing with experimental results.

9.2 DESIGN OF TEST SPUR GEAR PAIRS

In order to predict the effect of geometrical features of gear pair on dynamic behavior, three gear pairs are designed by adding backlash, altering the tooth profile, and changing the addendum (tip relief). The variations in these parameters are within the permissible range, which are commonly applied in industries. The levels of three geometrical factors considered for the present study are decided after through consultations with experts from the gear-manufacturing domain, and the same are given in Table 9.1. The levels of all other gear design factors are kept constant. 20° involute gear tooth systems and one ratio test spur gear pairs have been considered. SAE 8620 was used for the test gear pair.

9.2.1 SPUR GEAR-PAIR TEST 1

The usual addendum for a 20° full-depth involute is 1.0 times module, but this gear pair has an addendum of 1.1 times module.

TABLE 9.1
Variables of Spur Gear Pairs

Variables	Unit	Pair 1	Pair 2	Pair 3
Module	mm	3	3	3
Number of teeth	-	50	50	50
Angle of pressure	deg.	20°	20°	20°
Face width	mm	20	20	20
Distance of center	mm	150	150	150
Root diameter	mm	142	142	142
Base diameter	mm	142.645	142.645	142.645
PCD	mm	145	145	145
OCD	mm	157	157	157
Addendum	mm	4.3	4.0	4.3
Dedendum	mm	3.33	3.33	3.33
contact ratio	-	1.2	1.2	1.2
Depth of teeth	mm	6.69	6.69	6.69
Backlash	mm	0.10–0.12	0.15–0.17	0.10–0.12
Circular tooth thickness	mm	3.6–3.64	3.63–3.62	3.6–3.6
Linear tip relief				
a. Δ	μm	Original	Original	15–20
b. L	mm	Zero relief	Zero relief	Long relief
		(ZR)	(ZR)	(LR)
		$L=0$	$L=0$	$L=Ln$

9.2.2 Spur Gear-Pair Test 2

The usual value of backlash for a 3.00 mm module and a 150 mm center distance is 0.100–0.125 mm. This gear pair has backlash in the range of 0.150–0.175 mm [19].

9.2.3 Spur Gear-Pair Test 3

Long-tip relief and profile adjustment are present in this gear combination. The modification of a profile can be described by two variables [2]: first, the relief at the gear tooth's tip, $\Delta = 15$–20 micron, and second, modification length, $Ln = 2.647$ mm.

9.3 FINITE ELEMENT ANALYSIS OF TEST SPUR GEAR PAIRS

To determine the dynamic response of three test spur gear pairs, a simulation study using an FE analysis is performed at an operating speed of 1200 rpm and an external torque of magnitude 5.6 N-m. The procedure adopted for this simulation study is as follows.

9.3.1 Introduction

The FE analysis is divided into three stages: the first is pre-processing, the second is solution, and the third is post-processing. In pre-processing, a 3D model is prepared and related input data like properties are defined. Meshing is done considering the type of element and size. Boundary conditions are applied. In the second stage, i.e., solution, all equations are to be solved. The third stage includes post-processing, and the results are to be displayed.

9.3.2 Procedure of Simulation Study

The geometry of a gearbox comprises a set of gears, two shafts, and four bearings inside the casing. The input shaft of the gearbox is connected to a motor shaft for the motion, and the output shaft of the gearbox is connected to rope brake dynamometer, where the torque is applied. Further steps to carry out simulation study are discussed below.

9.3.2.1 Modeling of Gearbox

Modeling of gearbox is the first and most important step in the FE analysis. An effective and reliable modeling is necessary to predict vibratory response of gearbox within a sufficient level of accuracy. Therefore, in this study, modeling is prepared by including important features of gearbox. At most care was taken as the gearbox is a complex mechanical system, and it consists of several components such as gear-pair, shafts, bearings, and lower and upper halves of casing.

First, a CAD model for each component in gearbox assembly is prepared using CATIA software. The geometrical modification of gear teeth is performed in space claim software of ANSYS. The next step is pre-processing, which includes removal of unwanted geometry, consideration of domain which is important for the FE analysis, and making a geometry suitable for meshing or discretization. The geometry pre-processing takes longer time as it is essential to correct pin holes, fillets, chamfers, and step geometry created during the CAD development. This unwanted geometry will increase the elements and nodes in gearbox assembly, which affects the solution and solution time.

Figure 9.1a and b shows 3D CAD modeling of spur gearbox before and after pre-processing. From Figure 9.1b, it is seen that CAD modeling is more simple and without holes. Also, shaft coupling is removed because it does not have much importance in the FE analysis. Only input shaft is required to provide input rotation as the boundary condition. Then, in FE analysis software, material properties are defined for each component in gearbox assembly.

9.3.2.2 Meshing of Gearbox

The meshing for the project contains only tetrahedral elements for geometry and contact elements for the contact purpose. The tetrahedral elements are SOLID 187 and SOLID 186, and for contact, elements CONTA 174 and TARGE 170 are used. The complex geometry of gear and gearbox does not allow the tetrahedron elements, as they are more difficult to be sliced in different geometries with simple dimensions. So, tetrahedral elements are preferred which capture the maximum part of the smallest portion of the gear. Also, in the FE analysis, load on one gear should be transferred to the mating gear, which needs to define the contact and target elements.

The SOLID 187 is a 3-D, 10 nodes element, whereas, SOLID 186 is a higher order 3-D, 20-nodes element. These elements have a quadratic displacement behavior and are well suited to modeling irregular meshes. The elements mostly suited for the sturdy geometry comprise chamfers, fillets, and pin holes. For the geometries that cannot be sliced to make simple geometry structures, tetrahedral element is the best solution.

These elements have three degrees of freedom at each node in the x, y, and z translational directions. Plasticity, hyperelasticity, creep, stress stiffening, big deflection,

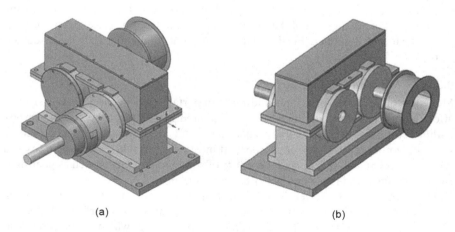

(a)   (b)

FIGURE 9.1   3D CAD modeling pre-processing (a) before and (b) after.

and large strain are all capabilities of these elements. It simulates the deformation of fully incompressible elastoplastic materials with incompressible hyperelastic materials based on mixed formulation technology.

The surface of one body is considered the contact surface, and the surface of the other body is the target surface while analyzing the interaction between two bodies. The target and contact surface constitute a contact pair. It is feasible to simulate straight and curved surfaces in 3D using these surface-to-surface contact elements. Frequently, simple geometric shapes like circles, parabolas, spheres, cones, and cylinders are used. Additionally, more intricate rigid structures or general deformable forms are modeled using specialized methods.

The number of elements created in the meshing of gearbox is 79,988, in which 1960 elements are SOLID 186 and the rest are SOLID 187. The number of nodes created in the meshing is 1,52,703. Figures 9.2 and 9.3 show meshing of gearbox and gears.

9.3.2.3   Assembly of Elements

Assembly of elements produces a set of equations that describe the structure. An FE structure is an assembly of elements. The behavior of a geometric model is characterized or approximated by specific differential equations and boundary conditions; it becomes a mathematical model. This step results in a matrix equation, which is called the FE model. The resulting algebraic equations will have more unknowns than algebraic equations since the elements are connected to neighbors. To eliminate more unknowns, the components need to be assembled.

FIGURE 9.2   Meshing of gearbox.

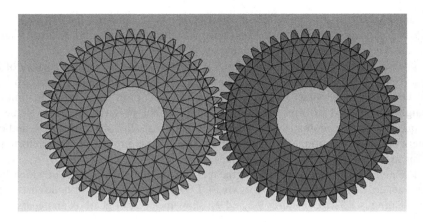

FIGURE 9.3 Meshing for gear pair.

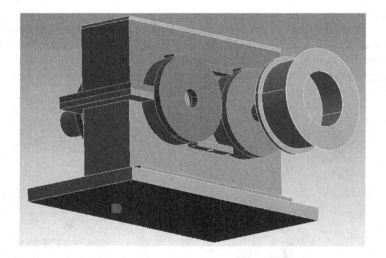

FIGURE 9.4 Fixed boundary condition for the modal analysis.

9.3.2.4 Boundary Conditions

The next step is to impose the boundary conditions to the assembly of gearbox. Also, material properties are provided to each part in the assembly of gearbox. The boundary conditions used in the simulation are depending on the experimental setup.

For a modal analysis, fixed support boundary condition is used to calculate the mode shape of the structure of the gearbox. As the base of gearbox was fixed in the experimental setup, a fixed support condition is used in the simulation. Figure 9.4 shows the fixed boundary condition given for gearbox.

And, for random analysis, PSD (power spectral density) acceleration is used as these were the data taken from the accelerometer. This PSD acceleration is provided at the base where fixed support was given. The random vibration input to the analysis is given in the form of PSD data only, which is the square of Fourier transform for a specific limited range of frequency.

9.3.2.5   Analysis

After performing meshing and imposing boundary conditions to the assembly of gearbox model, the next important step is finding the solution of equations. Certain assumptions are needed to idealize the physical issue into a mathematical model, which results in differential equations governing the mathematical model.

The FE method often transforms the discrete continuum model's differential equations into a set of algebraic equations. Once boundary conditions are set, the corresponding algebraic equations are computed for the solution at the mesh points. The discrete equations associated with the FE mesh are solved with the help of an equation solver. The choice of equation solver is important for large and nonlinear problems. The two basic solvers are used: one is Block Lanczos solver for modal analysis, and the other is random vibration solver.

9.3.2.6   Results

The result of root-mean-square (RMS) acceleration (g) value is taken at the position, where accelerometer has been placed. The directional RMS acceleration (g) in the y-direction is taken for all bodies included in the assembly. The RMS acceleration (g) value is the only output to be seen from the random vibration analysis.

For the analysis purpose, all three tests were conducted at T = 5.6 N-m torque and speed (N = 1200 rpm) condition. The sample frequency graph gained from the simulation study for gear-pair number 1 is shown in Figure 9.5. The responses obtained from the FE analysis study for gear pairs 1–3 are shown in Table 9.2.

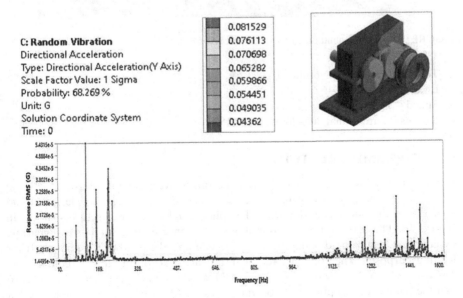

FIGURE 9.5   RMS acceleration (g) versus frequency plot for gear-pair number 1.

TABLE 9.2

Responses of Three Test Spur Gear Pairs Using FE Analysis

Pairs	Designation	RMS Acceleration (g)
1	Gear model 1	0.081
2	Gear model 2	0.084
3	Gear model 3	0.103

FIGURE 9.6   Experimental gear dynamic test setup.

Test Spur Gear-Pair Number 1
Loading condition – 5.6 N-m
Speed – 1200 rpm
Results – Directional Acceleration (RMS)

9.4   EXPERIMENTAL STUDY

In order to validate FE analysis results, experiments were conducted using the same test spur gear pairs. For that purpose, gear dynamic test setup was designed and constructed to collect vibration data. The photograph of the test setup is shown in Figure 9.6. The drive (5 hp induction motor) is connected to a single-stage gearbox by coupling. The rope dynamometer is connected to the gearbox's output shaft for loading purposes. The applied torque can be varied according to the requirement. The drive of novel test setup is equipped with an electronic speed control device, so that the input speed of pinion can be varied easily. All three tests were conducted at an operating speed of 1200 rpm and an external torque of magnitude 5.6 N-m.

9.4.1 PROCEDURE TO CONDUCT EXPERIMENTS

To study the effect of variation in geometrical features of spur gear pairs on vibrations, the experiments were conducted, wherein the rms acceleration amplitudes were measured. An accelerometer (Make: CTC, Model: AC102-1A, Sensitivity: 100 mv/g) was mounted on the location in the direction identified through sweep test, i.e. bearing block number 4 along the vertical direction. A microphone was used to measure the noise level of gearbox in terms of decibel (dB). The vibration signals, particularly in frequency and time domains, are captured and analyzed in real time using a fast Fourier transform (FFT) analyzer. For further analysis, only frequency-domain RMS acceleration (g) is used. The FFT analyzer is used for analyzing actual vibration.

The vibration signals were measured at a sampling rate of 4096 Hz, stored in a tape-recorder, and consequently analyzed in the FFT analyzer. The frequency range was chosen to be 5–1000 Hz in order to seizure mesh frequency. A bandwidth of 1 Hz was chosen for sideband leakage for experimental spectra analysis. The time length of signal was chosen for 1 second. It is dependent on the frequency span and the number of analyzer lines used. In this experiment, 1600 analyzer lines were used to display 4096 point transform. After filtering, the RMS value of the raw observed vibration signal is used to determine the amplitude of the reference signal. The final link in the measurement chain before data display is the average. To eliminate noise and unwanted random sources, the frequency spectra were averaged using four linear averages. The frequency graph of pair 1 is shown in Figure 9.7.

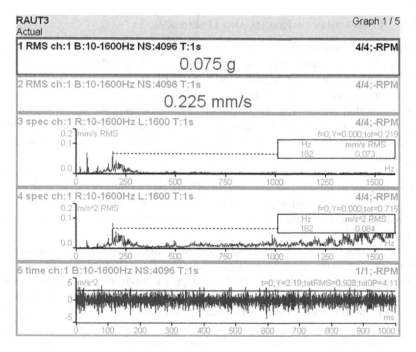

FIGURE 9.7   Frequency graph of pair 1.

TABLE 9.3

Samples of Gear Pairs

Pairs	Designation	RMS Acceleration (g)
1	Gear model 1	0.065
2	Gear model 2	0.073
3	Gear model 3	0.094

TABLE 9.4

Summary of Experimental and FE Analysis Results

Pairs	Testing	Experimental Responses RMS Acceleration (g)	FEA Responses RMS Acceleration (g)
1	Gear model 1	0.065	0.071
2	Gear model 2	0.073	0.084
3	Gear model 3	0.094	0.103

Table 9.3 displays the performance of three test spur gear pairs for changes in the addendum, backlash, and profile in terms of maximum acceleration (g).

9.4.2 COMPARISON OF RESULTS AND DISCUSSION

For validation purpose, the responses obtained from the FE analysis study are compared with the experimental results and are reproduced in Table 9.4.

From Table 9.4, it can be seen that the trend of FE analysis results is similar to that of experimental results. The responses obtained from the FE analysis is 10%–15% greater than that of experimental responses. This is because the actual working conditions are slightly different from those of the simulation studies. From these studies, it is observed that a slight change in one geometrical feature of gear pair will result in the change of dynamic responses.

From this experimental and FE analysis study, it is observed that gear tooth addendum, backlash, and tip relief have effects on the dynamics of gear pair. All these parameters are nonlinear in nature. These studies were performed by varying only one parameter, while the other gear parameters and operating conditions are held constants.

From the results, it is observed that gear-pair number 3 shows the maximum acceleration level as compared with gear-pair numbers 1 and 2. The maximum dynamic response in gear-pair number 3 is due to tooth profile modification. As a consequence of the findings, it is seen that the linear long-tip relief is the first significant component in the modification of the tooth profile. Excessive profile modification reduces the contact ratio, which increases the dynamic load resulting in higher vibrations [2,20]. Gear backlash and tooth addendum are the second and third dominant factors, respectively.

9.5   CONCLUSION

In the current study, an effort has been made to examine the impacts of several factors—including addendum, backlash, and profile alteration—arising from a modification in the geometrical profile of the gear pair on the dynamic response of the gear pair. According to the observations made about linear tip-relief tooth profile adjustment, backlash and addendum are the two factors that contribute least to generation of vibration. Therefore, it can be inferred from the current research that the dynamic response of spur gear pairs can be significantly influenced by the best combination of geometrical characteristics and their values. Additionally, it is observed that the geometrical characteristics of the gear tooth profile can be chosen in a way that minimizes vibration that is produced during operation.

REFERENCES

[1] Kahraman, A., Ozguven, H. N., Houser, D. R., and Zakrajsek, J. J., 1992. Dynamic analysis of geared rotors by finite elements. *ASME Journal of Mechanical Design,* Vol. 114, No. 3, pp. 507–514.

[2] Lin, H. H., Oswald, F. B., and Townsend, D. P., 1994. Dynamic loading of spur Gears with linear or parabolic tooth profile modifications. *Mechanism and Machine Theory,* Vol. 29, No. 8, pp. 1115–1129.

[3] Velex, P., and Maatar, M., 1996. A mathematical model for analyzing the influence of shape deviations and mounting errors on gear dynamic behavior. *Journal of Sound and Vibration,* Vol. 191, No. 5, pp. 629–660.

[4] Liou, C. H., Lin, H. H., Oswald, F. B., and Townsend, D. P., 1996. Effect of contact ratio on spur gear dynamic load with no tooth profile modification. *ASME Journal of Mechanical Design,* Vol. 118, No. 3, pp. 439–443.

[5] Ozguven, H. N., and Houser, D. R., 1988. Mathamatical models used in gear dynamics-a review. *Journal of Sound and Vibration,* Vol. 121, No. 3, pp. 383–411.

[6] Wang, J., Li, R., and Peng, X., 2003. Survey of nonlinear vibration of gear transmission systems. *ASME Journal of Applied Mechanics Review,* Vol. 56, No. 3, pp. 309–329.

[7] Wen, Y., Yang, J., and Wang, S., 2014. Random dynamics of a nonlinear spur gear pair in probabilistic domain. *Journal of Sound and Vibration,* Vol. 333, No. 20, pp. 5030–5041.

[8] Kabur, M., Kahraman, A., Zini, D. M., and Kienz, K., 2004. Dynamic analysis of a multi-shaft helical gear transmission by finite elements. *ASME Journal of Vibration and Acoustics,* Vol. 124, pp. 398–406.

[9] Maliha, R., Dogruer, C. U., and Ozguven, H. N., 2004. Nonlinear dynamic modeling of gear-shaft- disk-bearing systems using finite elements and describing functions. *ASME Journal of Mechanical Design, Vol.* 126, pp. 534–541.

[10] Ozguven, H. N., and Houser, D. R., 1988. Dynamic analysis of high speed gears by using loaded static transmission error. *Journal of Sound and Vibration,* Vol. 125, No. 1, pp. 71–83.

[11] Tamminana, V. K., Kahraman, A., and Vijaykumar, S., 2007. A study of the relationship between the dynamic factors and the dynamic transmission error of spur gear pairs. *ASME Journal of Mechanical Design,* Vol. 129, pp. 75–84.

[12] Li, S., 2007. Effects of machining errors, assembly errors and tooth modifications on loading capacity, load-sharing ratio and transmission error of a pair of spur gears. *Mechanism and Machine Theory, Vol.* 42, No. 6, pp. 698–726.

[13] Kim, W., Yoo, H. H., and Chung, J., 2010. Dynamic analysis for a pair of spur gears with translational motion due to bearing deformation. *Journal of Sound and Vibration,* Vol. 329, No. 21, pp. 4409–4421.

[14] Moradi, H., and Salarieh, H., 2012, Analysis of nonlinear oscillations in spur gear pairs with approximated modeling of backlash nonlinearity. *Machine and Mechanism Theory,* Vol. 51, pp. 14–31.

[15] Ma, H., Pang, X., Feng, R., and Wen, B., 2016. Evaluation of optimum profile modification curves of profile shifted spur gears based on vibration responses. *Mechanical Systems and Signal Processing,* Vol. 70–71, pp. 1131–1149.

[16] Wang, J., He, G., Zhang, J., Zhao, Y., and Yao, Y., 2017. Nonlinear dynamic analysis of Spur gear system for railway locomotive. *Mechanical Systems and Signal Processing,* Vol. 85, pp. 41–55.

[17] Liu, P., Zhu, L., Gou, X., Shi, J., and Lin, G., 2021. Dynamic modeling and analyzing of Spur gear pair with pitch deviation considering time varying contact ratio under multi-state meshing. *Journal of Sound and Vibration,* Vol. 513, pp. 116411.

[18] Gao, P., Liu, H., Yan, P., Xie, Y., Xiang, C., and Wang, C., 2022. Research on application of dynamic optimization modification for an involute spur gear in a fixed-shaft gear transmission system, *Mechanical Systems and Signal Processing,* Vol. 181, pp. 109530.

[19] Maitra, G. M., 1994, *Handbook of Gear Design,* Second Edition, Tata McGraw-Hill Publishing Company Limited, New Delhi.

[20] Ratanasumawong, C., Matsumura, S., Tatsuno, T., and Houjoh, H., 2009. Estimating gear tooth surface geometry by means of the vibration characteristics of gears with tooth surface form error. *ASME Journal of Mechanical Design,* Vol. 131, No. 10, pp. 1–9.

Index

accelerometer 3–5, 16, 90, 91,93
active vibration control 51, 52, 61
actuator 41, 51–54, 59, 61, 71

base isolation 47, 49
blind source separation 13, 14, 21, 24
boundary conditions 52, 87, 89–91
building 32, 33, 35–39, 42, 50, 65

cantilever beam 51, 53, 54, 61
controller 27, 29, 31, 51, 52, 55–61
cylinder 33, 34, 89

damping 3, 32–37, 39, 42–44, 50, 72, 82, 83
Daubechies wavelet 11, 14
dynamic load 84, 85, 94, 95
dynamic response 22, 50, 85, 87, 94, 95

earthquake 33, 37–42, 44, 46–50
empirical mode decomposition 13, 15
energy 11, 12 , 22, 25, 32, 33, 39, 41, 51, 54, 82

fast Fourier transform 2, 77, 93
feedback 25, 26, 28, 29, 31, 55–61, 71, 72
flexible 3, 25, 26, 27, 29, 31, 41, 61, 66, 83

gear 2, 11, 43, 84–96
gearbox 2, 87–93
genetic algorithm 51–55, 57, 59, 61

Helmholtz 71–74, 76–79, 82, 83
hyperelastic 88, 89

instability 42, 71–74, 76–80, 82, 83
Internet of Things 10

journal bearings 1–5, 7, 9, 11

knurling 74

linear quadratic regulator 52
looseness 2, 4–7, 9, 10

MATLAB 2, 37, 40, 46, 53–59, 61
mitigation 49, 63, 65, 66

nonlinear 26, 83–85, 91, 94–96

oil whirl 10
optimal control 51, 52, 55, 57–61

passive control 37, 41, 83
pendulum 32–39
potential 12, 37, 71

quadratic cost function 55

reliable 32, 87
resonator 71–79, 82, 83
Riccati 27, 55–58
Rijke 71, 73–79, 82
Root-mean-square 37, 91

seismic 32, 33, 35–39, 41–45, 47, 49, 50
sensor 12, 13, 15–17, 19, 21, 22, 41, 51–54, 60, 61, 67, 75
shaft 1–6, 9, 85, 87, 88, 92, 95, 96
singular spectrum analysis 13, 24
sloshing 32, 39, 42, 44, 46, 47, 49
spectrum 2, 5–9, 13, 24, 63
state space 54, 57
structural health monitoring 12, 13, 15, 17, 19, 21–24

trajectory 13, 26, 29
transmission 57, 63, 65, 66, 84, 85, 95, 96

unsteady 71

vector 17, 19, 20, 42, 53–55, 57
vibration control 25, 27, 29, 31–33, 35, 37, 39, 41, 49–52, 60, 61, 70

wall 33, 35, 42, 44, 66, 67

Printed in the United States
by Baker & Taylor Publisher Services